Contents

2010 ARRL Field Day Rules	9
US Amateur Radio Band Chart	15
The Considerate Operator's Frequency Guide	16
ARRL/RAC Section Abbreviations	17
ARRL/RAC Section Map	18
Morse Code Character Set	19

Antenna Projects from the Pages of *QST*

Having a Field Day with the Moxon Rectangle L.B. Cebik, W4RNL	21
An Easily Constructed 30- and 40-Meter Trap Dipole Antenna Arthur Gillespie Jr, K4TP	26
A Simple HF-Portable Antenna Phil Salas, AD5X	27
A Ground Coupled Portable Antenna Robert Johns, W3JIP	29
The NJQRP Squirt Joe Everhart, N2CX	34
Gain without Pain – A Beam Antenna for Field Day Bob Clarke, N1RC	38
The Arkansas Catfish Dipole David G. Byrd, KD7VA	42
A Three Element Lightweight Monobander for 14 MHz David Reid, PA3HBB	44
A Simple Fixed Antenna for VHF/UHF Satellite Work L.B. Cebik, W4RNL	48
A Portable 2-Element Triband Yagi Markus Hansen, VE7CA	52
A Portable Twin-Lead 20-Meter Dipole Rich Wadsworth, KF6QKI	55
A "One-Masted Sloop" for 40, 20, 15 and 10 Meters Rick Rogers, KI8GX	57
One Stealthy Delta Steve Ford, WB8IMY	60
A Horizontal Loop for 80-Meter DX John S. Belrose, VE2CV	62

Tips for Field Day Power

    Power Packing for Emergencies   John S. Raydo, K0IZ   **69**

    Modern Portable Power Generators – Small, Sleek and Super Stable!   **72**
        Kirk Kleinschmidt, NT0Z

Public Relations

    Press Release   **77**

    Publicity Tip Sheet   **78**

Amateur Satellites   Steve Ford, WB8IMY   **79**

HF Digital Communications   Steve Ford, WB8IMY   **85**

References

    RF Connectors and Transmission Lines   **93**

    US Customary Units and Conversion Factors   **99**

    Voltage-Power Conversion Table   **101**

    Reflection Coefficient, Attenuation, SWR and Return Loss   **102**

    Abbreviations   **103**

    Antenna and Tower Safety   **106**

2010 Field Day Log   **114**

Foreword

Field Day is, without question, the largest on-air event in Amateur Radio. Its roots are in the efforts of hams to exercise their emergency communication skills, starting with the first Field Day many decades ago. Emergency preparedness is still at the core of Field Day, although some also see it as a competition while others celebrate it as a social event. Since the beginning, Field Day has encompassed all of these things!

Regardless of how you participate in Field Day, I believe you'll find this book to be a valuable reference. The antenna projects alone will spark your imagination. You'll also learn how to add to your Field Day point totals by making satellite contacts and using digital modes. Finally, this book includes a handy Field Day log.

Be safe and have fun!

73,

Dave Sumner, K1ZZ
ARRL Chief Executive Officer

About the ARRL

The seed for Amateur Radio was planted in the 1890s, when Guglielmo Marconi began his experiments in wireless telegraphy. Soon he was joined by dozens, then hundreds, of others who were enthusiastic about sending and receiving messages through the air—some with a commercial interest, but others solely out of a love for this new communications medium. The United States government began licensing Amateur Radio operators in 1912.

By 1914, there were thousands of Amateur Radio operators—hams—in the United States. Hiram Percy Maxim, a leading Hartford, Connecticut inventor and industrialist, saw the need for an organization to band together this fledgling group of radio experimenters. In May 1914 he founded the American Radio Relay League (ARRL) to meet that need.

Today ARRL, with approximately 155,000 members, is the largest organization of radio amateurs in the United States. The ARRL is a not-for-profit organization that:
• promotes interest in Amateur Radio communications and experimentation
• represents US radio amateurs in legislative matters, and
• maintains fraternalism and a high standard of conduct among Amateur Radio operators.

At ARRL headquarters in the Hartford suburb of Newington, the staff helps serve the needs of members. ARRL is also International Secretariat for the International Amateur Radio Union, which is made up of similar societies in 150 countries around the world.

ARRL publishes the monthly journal *QST*, as well as newsletters and many publications covering all aspects of Amateur Radio. Its headquarters station, W1AW, transmits bulletins of interest to radio amateurs and Morse code practice sessions. The ARRL also coordinates an extensive field organization, which includes volunteers who provide technical information and other support services for radio amateurs as well as communications for public-service activities. In addition, ARRL represents US amateurs with the Federal Communications Commission and other government agencies in the US and abroad.

Membership in ARRL means much more than receiving *QST* each month. In addition to the services already described, ARRL offers membership services on a personal level, such as the Technical Information Service—where members can get answers by phone, email or the ARRL website, to all their technical and operating questions.

Full ARRL membership (available only to licensed radio amateurs) gives you a voice in how the affairs of the organization are governed. ARRL policy is set by a Board of Directors (one from each of 15 Divisions). Each year, one-third of the ARRL Board of Directors stands for election by the full members they represent. The day-to-day operation of ARRL HQ is managed by an Executive Vice President and his staff.

No matter what aspect of Amateur Radio attracts you, ARRL membership is relevant and important. There would be no Amateur Radio as we know it today were it not for the ARRL. We would be happy to welcome you as a member! (An Amateur Radio license is not required for Associate Membership.) For more information about ARRL and answers to any questions you may have about Amateur Radio, write or call:

ARRL—The national association for Amateur Radio
225 Main Street
Newington CT 06111-1494
Voice: 860-594-0200
   Fax: 860-594-0259
   E-mail: **hq@arrl.org**
   Internet: **www.arrl.org/**

Prospective new amateurs call (toll-free):
800-32-NEW HAM (800-326-3942)
You can also contact us via e-mail at **newham@arrl.org**
or check out *ARRLWeb* at **www.arrl.org/**

Join ARRL and experience the BEST of Ham Radio!

ARRL Membership Benefits and Services:
- *QST* magazine — your monthly source of news, easy-to-read product reviews, and features for new hams!
- Technical Information Service — access to problem-solving experts!
- Members-only Web services — find information fast, anytime!
- Best amateur radio books and software
- Public service and emergency communications training

I want to join ARRL.
Send me the FREE book I have selected
(choose one)
☐ The ARRL Repeater Directory
☐ The ARRL Emergency Communication Handbook

_____   _____
Name   Call Sign

_____
Street

_____   _____   _____
City   State   ZIP

Please check the appropriate one-year[1] rate:
☐ $39 in US.
☐ Age 21 or younger rate, $20 in US (see note*).
☐ Canada $49.
☐ Elsewhere $62.

[1] 1-year membership dues include $15 for a 1-year subscription to QST. International 1-year rates include a $10 surcharge for surface delivery to Canada and a $23 surcharge for air delivery to other countries. Other US membership options available: Blind, Life, and QST by First Class postage. Contact ARRL for details.
*Age 21 or younger rate applies only if you are the oldest licensed amateur in your household.
International membership is available with an annual CD-ROM option (no monthly receipt of QST). Contact ARRL for details.
Dues subject to change without notice.

Sign up my family members, residing at the same address, as ARRL members too! They'll each pay only $8 for a year's membership, have access to ARRL benefits and services (except QST) and also receive a membership card.

☐ Sign up _____ family members @ $8 each = $ _____. Attach their names & call signs (if any).

☐ Total amount enclosed, payable to ARRL $ _____ . (US funds drawn on a bank in the US).
☐ Enclosed is $ _____ ($1.00 minimum) as a donation to the Legal Research and Resource Fund.
☐ Charge to: ☐ VISA   ☐ MasterCard   ☐ Amex   ☐ Discover

_____   _____
Card Number   Expiration Date

_____
Cardholder's Signature

Call Toll-Free (US) **1-888-277-5289**
Join Online **www.arrl.org/join** or
Clip and send to:

225 Main Street
Newington, CT 06111-1494 USA

☐ If you do not want your name and address made available for non-ARRL related mailings, please check here.

FD HBK

ARRL Field Day 2010 Rules

1. Eligibility: Field Day is open to all amateurs in the areas covered by the ARRL/RAC Field Organizations and countries within IARU Region 2. DX stations residing in other regions may be contacted for credit, but are not eligible to submit entries.

2. Object: To work as many stations as possible on any and all amateur bands (excluding the 60, 30, 17, and 12-meter bands) and in doing so to learn to operate in abnormal situations in less than optimal conditions. A premium is placed on developing skills to meet the challenges of emergency preparedness as well as to acquaint the general public with the capabilities of Amateur Radio.

3. Date and Time Period: Field Day is **always the fourth full weekend of June**, beginning at 1800 UTC Saturday and ending at 2100 UTC Sunday. **Field Day 2010 will be held June 26-27, 2010.**

    3.1. Class A and B (see below) stations that do not begin setting up until 1800 UTC on Saturday may operate the entire 27-hour Field Day period.

    3.2. Stations who begin setting up before 1800 UTC Saturday may work only 24 consecutive hours, commencing when on-the-air operations begin.

    3.3. No Class A or B station may begin their set-up earlier than 1800 UTC on the Friday preceding the Field Day period.

4. Entry Categories: Field Day entries are classified according to the maximum number of *simultaneously* transmitted signals, followed by a designator indicating the nature of their individual or group participation. Twenty (20) transmitters maximum are eligible for the purpose of calculating bonus points (2,000 points maximum). However, additional transmitters may be used simultaneously in determining your entry category. Switching and simulcasting devices are prohibited. **Bonus stations, such as the GOTA station and satellite station do not count towards determining the number of transmitters for the class and do not qualify for transmitter bonus points.**

    4.1. (Class A) Club / non-club portable: Club or a non-club group of three or more persons set up specifically for Field Day. Such stations must be located in places that are not regular station locations and must not use facilities installed for permanent station use, or use any structure installed permanently for Field Day. A single licensee or trustee for the entry is responsible for the group entry. All equipment (including antennas) must lie within a circle whose diameter does not exceed 300 meters (1000 feet). To be listed as Class A, all contacts must be made with transmitter(s) and receiver(s) operating independent of commercial power mains. Entrants whom for any reason operate a transmitter or receiver from a commercial main for one or more contacts will be listed separately as Class A-Commercial.

        4.1.1. Get-On-The-Air (GOTA) Station. Any Class A (or F) entry whose transmitter classification is two or more transmitters may also operate one additional station without changing its base entry category, known as the GET-ON-THE-AIR (GOTA) station. **This GOTA station may operate on any Field Day band, HF or VHF, but is limited to one transmitted signal at any time.**

            4.1.1.1. This station **must use** a different callsign from the primary Field Day station. The GOTA station must use the same callsign for the duration of the event regardless if operators change. **The GOTA station uses the same exchange as its parent.**

            4.1.1.2. **The GOTA station may be operated by any person licensed since the previous year's Field Day, regardless of license class. It may also be operated by a generally inactive licensee. Non-licensed persons may participate under the direct supervision of an appropriate control operator. A list of operators and participants must be included on the required summary sheet to ARRL HQ.**

            4.1.1.3. **As per FCC rules, this station must have a valid control operator present if operating beyond the license privileges of the participant using the station.**

            4.1.1.4. The maximum transmitter output power for the GOTA station shall be **150 watts**. If the primary Field Day group is claiming the QRP multiplier level of 5, the maximum transmitter output power of the GOTA station may not exceed 5 watts.

            4.1.1.5. A maximum of 500 QSOs made by this station may be claimed for credit by its primary Field Day operation. In addition, bonus points may be earned by this station under rule 7.3.13.

            4.1.1.6. The GOTA station may operate on any Field Day band. Only one transmitted signal is allowed from the GOTA station at any time.

4.1.1.7. The GOTA station does not affect the additional VHF/UHF station provided for under Field Day Rule 4.1.2. for Class A stations.

4.1.1.8. Participants are reminded that non-licensed participants working under the direction of a valid control operator may only communicate with other W/VE stations or with stations in countries with which the US has entered a third-party agreement.

4.1.1.9. The GOTA station does not qualify as an additional transmitter when determining the number of transmitters eligible for the 100-point emergency power bonus under Rule 7.3.1.

4.1.2. **Free VHF Station:** Any Class A entry whose category is two or more transmitters may also operate **one additional transmitter** if it operates exclusively on any band or combination of bands above 50 MHz (VHF/UHF) without changing its basic entry classification. **This station does not qualify for a 100-point bonus as an additional transmitter**. This station may be operated for the clubs Field Day period and all contacts count for QSO credit. It is operated using the primary callsign and exchange of the main Field Day group and is separate and distinct from the GOTA station.

4.2. (Class A - Battery) Club / non-club portable: Club or non-club group of three or more persons set up specifically for Field Day. All contacts must be made using an output power of 5 Watts or less **and** the power source must be something other than commercial power mains or motor-driven generator (e.g.: batteries, solar cells, water-driven generator). Other provisions are the same for regular Class A. Class AB is eligible for a GOTA station if GOTA requirements are met.

4.3. (Class B) One or two person portable: A Field Day station set up and operated by no more than two persons. Other provisions are the same for Class A except it is not eligible for a GOTA or free VHF station. One and two person Class B entries will be listed separately.

4.4. (Class B - Battery) One or two person portable: A Field Day station set up and operated by no more than two persons. All contacts must be made using an output power of 5 Watts or less **and** the power source must be something other than commercial mains or motor-driven generator. Other provisions are the same for Class A except it is not eligible for a GOTA or free VHF station. One and two person Class B - Battery entries will be listed separately.

4.5. (Class C) Mobile: Stations in vehicles capable of operating while in motion and normally operated in this manner. This includes maritime and aeronautical mobile. If the Class C station is being powered from a car battery or alternator, it qualifies for emergency power but does not qualify for the multiplier of 5, as the alternator/battery system constitutes a motor-driven generating system.

4.6. (Class D) Home stations: Stations operating from permanent or licensed station locations using commercial power. Class D stations may only count contacts made with Class A, B, C, E and F Field Day stations.

4.7. (Class E) Home stations - Emergency power: Same as Class D, but using emergency power for transmitters and receivers. Class E may work all Field Day stations.

4.8. (Class F) Emergency Operations Centers (EOC): An amateur radio station at an established EOC activated by a club or non-club group. Class F operation must take place at an established EOC site. Stations may utilize equipment and antennas temporarily or permanently installed at the EOC for the event. Entries will be reported according to number of transmitters in simultaneous operation. Class F stations are eligible for a GOTA and free VHF station at Class 2F and above.

4.8.1. For Field Day purposes, an Emergency Operations Center (EOC) is defined as a facility established by:

a) a Federal, State, County, City or other Civil Government, agency or administrative entity; or,

b) a Chapter of a national or international served agency (such as American Red Cross or Salvation Army) with which your local group has an established operating arrangement;

4.8.1.1. A private company EOC does not qualify for Class F status unless approved.

4.8.2. Planning of a Class F operation must take place in conjunction and cooperation with the staff of the EOC being activated.

4.8.3. Other provisions not covered are the same as Class A.

4.8.4. A Class F station may claim the emergency power bonus if emergency power is available at the EOC site.

4.8.4.1. The emergency power source must be tested during the Field Day period but you are not required to run the Class F operation under emergency power.

5. Exchange: Stations in ARRL / RAC sections will exchange their Field Day operating Class and ARRL / RAC section. Example: a three transmitter class A station in Connecticut which also has a GOTA station and the extra

VHF station would send "3A CT" on CW or "3 Alpha Connecticut" on Phone. DX stations send operating class and the term DX (i.e. 2A DX).

6. Miscellaneous Rules:

6.1. A person may not contact for QSO credit any station from which they also participate.

6.2. A transmitter/receiver/transceiver used to contact one or more Field Day stations may not subsequently be used under any other callsign to participate in Field Day. Family stations are exempt provided the subsequent callsign used is issued to and used by a different family member.

6.3. Phone, CW and Digital (non-CW) modes on a band are considered as separate bands. A station may be worked only once per band under this rule.

6.4. All voice contacts are equivalent.

6.5. All non-CW digital contacts are equivalent.

6.6. Cross-band contacts are not permitted (Satellite QSOs cross-band contacts are exempted).

6.7. The use of more than one transmitter at the same time on a single band-mode is prohibited. Exception: a dedicated GOTA station may operate as prescribed in Rule 4.1.

6.8. No repeater contacts are allowed.

6.9. Batteries may be charged while in use. Except for Class D stations, the batteries must be charged from a power source other than commercial power mains. To claim the power multiplier of five, the batteries must be charged from something other than a motor driven generator or commercial mains.

6.10. All stations for a single entry must be operated under one callsign, except when a dedicated GOTA station is operated as provided under Field Day Rule 4.1.1. it uses a single, separate callsign.

7. Scoring: Scores are based on the total number of QSO points times the power multiplier corresponding to the highest power level under which any contact was made during the Field Day period plus the bonus points.

7.1. QSO Points:

7.1.1. Phone contacts count one point each.

7.1.2. CW contacts count two points each.

7.1.3. Digital contacts count two points each.

7.2. Power multipliers: The power multiplier that applies is determined by the highest power output of any of the transmitters used during the Field Day operation.

7.2.1. If all contacts are made using a power of 5 Watts or less *and* if a power source other than commercial mains or motor-driven generator is used (batteries, solar cells, water-driven generator), the power multiplier is 5 (five).

7.2.2. If all contacts are made using a power of 5 Watts or less, but the power source is from a commercial main or from a motor-driven generator, the power multiplier is 2. If batteries are charged during the Field Day period using commercial mains or a motor-driven generator the power multiplier is 2 (two).

7.2.3. If any or all contacts are made using an output power up to 150 Watts or less, the power multiplier is 2 (two).

7.2.4. If any or all contacts are made using an output power greater than 150 Watts, the power multiplier is 1 (one).

7.2.5. The power multiplier for an entry is determined by the maximum output power used by any transmitter used to complete any contact during the event. (Example: a group has one QRP station running 3 Watts and a second station running 100 Watts, the power multiplier of 2 applies to all contacts made by the entire operation).

7.3. Bonus Points: All stations are eligible for certain bonus points, depending on their entry class. The following bonus points will be added to the score, after the multiplier is applied, to determine the final Field Day score. Bonus points will be applied only when the claim is made on the summary sheet and any proof required accompanies the entry or is received via email or normal mail delivery.

7.3.1. 100% Emergency Power: 100 points per transmitter classification if all contacts are made only using an emergency power source up to a total of 20 transmitters **(maximum 2,000 points.) GOTA station and free VHF Station for Class A and F entries do not qualify for bonus point credit and should not be included in the club's transmitter total.** All transmitting equipment at the site must operate from a power source completely independent of the commercial power mains to qualify. (Example: a club operating 3 transmitters plus a GOTA station and using 100% emergency power receives 300 bonus points.) **Available to Classes A, B, C, E, and F.**

7.3.2. Media Publicity: 100 bonus points may be earned for attempting to obtain publicity from the local media. A copy of the press release, or a copy of the actual media publicity received (newspaper article, etc) must be submitted to claim the points. **Available to all Classes.**

7.3.3. Public Location: 100 bonus points for physically locating the Field Day operation in a public place (i.e. shopping center, park, school campus, etc). The intent is for amateur radio to be on display to the public. **Available to Classes A, B and F.**

7.3.4. Public Information Table: 100 bonus points for a Public Information Table at the Field Day site. The purpose is to make appropriate handouts and information available to the visiting public at the site. A copy of a visitor's log, copies of club handouts or photos is sufficient evidence for claiming this bonus. **Available to Classes A, B and F.**

7.3.5. Message Origination to Section Manager: 100 bonus points for origination of a National Traffic System (NTS) style formal message to the ARRL Section Manager or Section Emergency Coordinator by your group from its site. You should include the club name, number of participants, Field Day location, and number of ARES operators involved with your station. The message must be transmitted during the Field Day period and a fully serviced copy of it must be included in your submission, in standard ARRL NTS format, or no credit will be given. The Section Manager message is separate from the messages handled in Rule 7.3.6. and may not be claimed for bonus points under that rule. **Available to all Classes.**

7.3.6. Message Handling: 10 points for each formal NTS style originated, relayed or received and delivered during the Field Day period, up to a maximum of 100 points (ten messages). Properly serviced copies of each message must be included with the Field Day report. **The message to the ARRL SM or SEC under Rule 7.3.6. does not count towards the total of 10 for this bonus. Available to all Classes. All NTS messages claimed for bonus points must leave or enter the site via amateur radio RF.**

7.3.7. Satellite QSO: 100 bonus points for successfully completing at least one QSO via an amateur radio satellite during the Field Day period. "General Rules for All ARRL Contests" (Rule 3.7.2.), (the no-repeater QSO stipulation) is waived for satellite QSOs. Groups are allowed one dedicated satellite transmitter station without increasing their entry category. Satellite QSOs also count for regular QSO credit. Show them listed separately on the summary sheet as a separate "band." You do not receive an additional bonus for contacting different satellites, though the additional QSOs may be counted for QSO credit unless prohibited under Rule 7.3.7.1. The QSO must be between two Earth stations through a satellite. **Available to Classes A, B, and F.**

   7.3.7.1 Stations are limited to one (1) completed QSO on any single channel FM satellite.

7.3.8. Alternate Power: 100 bonus points for Field Day groups making a minimum of five QSOs without using power from commercial mains or petroleum driven generator. This means an "alternate" energy source of power, such as solar, wind, methane or water. This includes batteries charged by natural means (not dry cells). The natural power transmitter counts as an additional transmitter. If you do not wish to increase your operating category, you should take one of your other transmitters off the air while the natural power transmitter is in operation. A separate list of natural power QSOs should be submitted with your entry. **Available to Classes A, B, E, and F.**

7.3.9. W1AW Bulletin: 100 bonus points for copying the special Field Day bulletin transmitted by W1AW (or K6KPH) during its operating schedule during the Field Day weekend (listed in this rules announcement). An accurate copy of the message is required to be included in your Field Day submission. (Note: The Field Day bulletin must be copied via amateur radio. It will not be included in Internet bulletins sent out from Headquarters and will not be posted to Internet BBS sites.) **Available to all Classes.**

7.3.10. Educational activity bonus: One (1) 100-point bonus may be claimed if your Field Day operation includes a specific educational-related activity. The activity can be diverse and must be related to amateur radio. It must be some type of formal activity. It can be repeated during the Field Day period but only one bonus is earned. For more information consult the FAQ in the complete Field Day packet. **Available to Classes A & F entries and available clubs or groups operating from a club station in class D and E with 3 or more participants.**

7.3.11. Site Visitation by an elected governmental official: One (1) 100-point bonus may be claimed if your Field Day site is visited by an elected government official as the result of an invitation issued by your group. **Available to all Classes.**

7.3.12. Site Visitation by a representative of an agency: One (1) 100-point bonus may be claimed if your Field Day site is visited by a representative of an agency served by ARES in your local community (American Red Cross, Salvation Army, local Emergency Management, law enforcement, etc) as the result of an invitation issued by your group. ARRL officials (SM, SEC, DEC, EC, etc) do not qualify for this bonus. **Available to all Classes.**

7.3.13. GOTA Bonus. Class A and F stations operating a GOTA station may earn the following bonus points:

7.3.13.1. When a GOTA operator successfully completes 20 QSOs, they receive 20 bonus points. Upon reaching an additional 20 QSOs the same operator receives a second 20 bonus points, up to a maximum of 100 Bonus points per GOTA operator. An operator may make more than 100 QSOs but the QSOs over 100 do not qualify for an additional bonus.

7.3.13.1.1. Additional GOTA operators may earn the GOTA bonus points under this rule, up to the maximum of 500 bonus points. (Remember that there is a 500-QSO limit for the GOTA station. But no single GOTA operator may earn more than 100 of the GOTA bonus points except as provided in 7.3.13.2.)

7.3.13.1.2. A single GOTA operator must complete all 20 QSOs required before the bonus is earned. There is no "partial credit" for making only a portion of the 20 QSOs or "pooling" QSOs between operators.

7.3.13.2. If a GOTA station is supervised full-time by a GOTA Coach, the bonus points earned for each 20 QSOs completed under Rule 7.3.13.1 will be doubled.

7.3.13.2.1. The GOTA Coach supervises the operator of the station, doing such things as answering questions and talking them through contacts, but may not make QSOs or perform logging functions.

7.3.13.2.2. To qualify for this bonus, there must be a designated GOTA Coach present and supervising the GOTA station at all times it is being operated.

7.3.14. Web submission: A 50-point bonus may be claimed by a group submitting their Field Day entry via the www.b4h.net/cabforms web site. **Available to all Classes.**

7.3.15. Field Day Youth Participation:

7.3.15.1. A 20-point bonus (maximum of 100) may be earned by any Class A, C, D, E, or F group for each participant age 18 or younger at your Field Day operation that completes at least one QSO.

7.3.15.2. For a 1-person Class B station, a 20-point bonus is earned if the operator is age 18 or younger. For a 2-person Class B station, a 20-point bonus is earned for each operator age 18 or younger (maximum of 40 points.) Keep in mind that Class B is only a 1 or 2 person operation. This bonus does not allow the total number of participants in Class B to exceed 1 or 2.

8. Reporting:

8.1. Entries may be submitted to the ARRL in one of three ways:

8.1.1. Via Field Day Web Submission Applet site at www.b4h.net/cabforms/;

8.1.2. Via email to fieldday@arrl.org; or

8.1.3. Via land postal or delivery service to Field Day Entries, 225 Main St, Newington, CT 06111.

8.2. Entries must be postmarked, emailed or submitted by July 27, 2010. Late entries cannot be accepted.

8.3. A complete Field Day Web Applet Submission site entry consists of:

8.3.1. An official ARRL summary sheet which is completed on the site;

8.3.2. Supporting information must be emailed to fieldday@arrl.org or submitted by land service. Supporting information must include:

8.3.2.1. An attached list of stations worked by band/mode during the Field Day period (dupe sheet or an alpha/numeric list sorted by band and mode); and

8.3.2.2. Proof of all bonus points claimed (copies of visitor logs, press releases, NTS messages handled, photographs, etc).

8.4. A complete non-applet email submission consists of:

8.4.1. An electronic copy of an ARRL summary sheet completely and accurately filled out;

8.4.2. An attached list of stations worked by band/mode during the Field Day period (dupe sheet or an alpha/numeric list sorted by band and mode); and

8.4.3. Proofs of bonus points claimed (copies of visitor logs, press releases, NTS messages handled, photographs, etc).

8.5. A complete land postal or delivery non-electronic submission consists of:

8.5.1. A complete and accurate ARRL summary sheet;

8.5.2. An accompanying list of stations worked by band/mode during the Field Day period (dupe sheet or an alpha/numeric list sorted by band and mode); and

8.5.3. Proofs of bonus points claimed (copies of visitor logs, press releases, NTS messages handled, photographs, etc).

8.6. Complete station logs are not required for submission. The club should maintain log files for one year in case they are requested by ARRL HQ.

8.7. Cabrillo format log files are not required for Field Day entries. They will be accepted in lieu of the dupe sheets but do not substitute for a summary sheet.

8.8. Digital images of proof of bonus points are acceptable.

8.9. Electronic submissions are considered signed when submitted.

9. Miscellaneous:

9.1. The schedule of bulletin times for W1AW is included in this announcement. While W1AW does not have regular bulletins on weekends, the Field Day message will be sent according to the schedule included with this announcement. The W1AW bulletins will be transmitted on the regular W1AW frequencies listed in *QST*. The PSK31 bulletin will be transmitted on the W1AW teleprinter frequencies.

9.1.2. The special Field Day bulletin will be transmitted from station K6KPH on the West Coast as included in the bulletin schedule.

9.2. See "General Rules for All ARRL Contests," "General Rules for All ARRL Contests on Bands Below 30 MHz," and "General Rules for All ARRL Contests on Bands Above 50 MHz" for additional rules (www.arrl.org/contests/forms) that may cover situations not covered in these Field Day rules.

9.3. Remember that the national simplex FM calling frequency of 146.52 MHz should not be used for making Field Day contacts.

9.4. The complete Field Day information package may be obtained by:

9.4.1. Sending a SASE with 5 units of postage to: Field Day Information Package, ARRL, 225 Main St., Newington, CT 06111; or

9.4.2. By downloading from the Contest Branch home page at: www.arrl.org/contests/forms

9.5. For more Field Day information/questions contact: fdinfo@arrl.org or phone (860) 594-0236.

Revised 2/2010

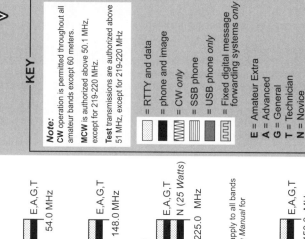

The Considerate Operator's Frequency Guide

The following frequencies are generally recognized for certain modes or activities (all frequencies are in MHz) during normal conditions. These are not regulations and occasionally a high level of activity, such as during a period of emergency response, DXpedition or contest, may result in stations operating outside these frequency ranges.

Nothing in the rules recognizes a net's, group's or any individual's special privilege to any specific frequency. Section 97.101(b) of the Rules states that "Each station licensee and each control operator must cooperate in selecting transmitting channels and in making the most effective use of the amateur service frequencies. No frequency will be assigned for the exclusive use of any station." No one "owns" a frequency.

It's good practice — and plain old common sense — for any operator, regardless of mode, to check to see if the frequency is in use prior to engaging operation. If you are there first, other operators should make an effort to protect you from interference to the extent possible, given that 100% interference-free operation is an unrealistic expectation in today's congested bands.

Frequencies	Modes/Activities	Frequencies	Modes/Activities
1.800-2.000	CW	18.100-18.105	RTTY/Data
1.800-1.810	Digital Modes	18.105-18.110	Automatically controlled data stations
1.810	CW QRP calling frequency	18.110	IBP/NCDXF beacons
1.843-2.000	SSB, SSTV and other wideband modes	21.060	QRP CW calling frequency
1.910	SSB QRP	21.070-21.110	RTTY/Data
1.995-2.000	Experimental	21.090-21.100	Automatically controlled data stations
1.999-2.000	Beacons	21.150	IBP/NCDXF beacons
		21.340	SSTV
3.500-3.510	CW DX window	21.385	QRP SSB calling frequency
3.560	QRP CW calling frequency		
3.570-3.600	RTTY/Data	24.920-24.925	RTTY/Data
3.585-3.600	Automatically controlled data stations	24.925-24.930	Automatically controlled data stations
3.590	RTTY/Data DX	24.930	IBP/NCDXF beacons
3.790-3.800	DX window		
3.845	SSTV	28.060	QRP CW calling frequency
3.885	AM calling frequency	28.070-28.120	RTTY/Data
3.985	QRP SSB calling frequency	28.120-28.189	Automatically controlled data stations
		28.190-28.225	Beacons
7.030	QRP CW calling frequency	28.200	IBP/NCDXF beacons
7.040	RTTY/Data DX	28.385	QRP SSB calling frequency
7.070-7.125	RTTY/Data	28.680	SSTV
7.100-7.105	Automatically controlled data stations	29.000-29.200	AM
7.171	SSTV	29.300-29.510	Satellite downlinks
7.285	QRP SSB calling frequency	29.520-29.580	Repeater inputs
7.290	AM calling frequency	29.600	FM simplex
		29.620-29.680	Repeater outputs
10.130-10.140	RTTY/Data		
10.140-10.150	Automatically controlled data stations		
14.060	QRP CW calling frequency		
14.070-14.095	RTTY/Data		
14.095-14.0995	Automatically controlled data stations		
14.100	IBP/NCDXF beacons		
14.1005-14.112	Automatically controlled data stations		
14.230	SSTV		
14.285	QRP SSB calling frequency		
14.286	AM calling frequency		

ARRL band plans for frequencies above 28.300 MHz are shown in *The ARRL Repeater Directory* and on **www.arrl.org**.

ARRL / RAC Section Abbreviations

1
Connecticut	CT	Rhode Island	RI
Eastern Massachusetts	EMA	Vermont	VT
Maine	ME	Western Massachusetts	WMA
New Hampshire	NH		

2
Eastern New York	ENY	Northern New York	NNY
NYC / Long Island	NLI	Southern New Jersey	SNJ
Northern New Jersey	NNJ	Western New York	WNY

3
Delaware	DE	Maryland – DC	MDC
Eastern Pennsylvania	EPA	Western Pennsylvania	WPA

4
Alabama	AL	Southern Florida	SFL
Georgia	GA	Tennessee	TN
Kentucky	KY	Virginia	VA
North Carolina	NC	West Central Florida	WCF
Northern Florida	NFL	Puerto Rico	PR
South Carolina	SC	US Virgin Islands	VI

5
Arkansas	AR	North Texas	NTX
Louisiana	LA	Oklahoma	OK
Mississippi	MS	South Texas	STX
New Mexico	NM	West Texas	WTX

6
East Bay	EB	San Diego	SDG
Los Angeles	LAX	San Francisco	SF
Orange	ORG	San Joaquin Valley	SJV
Santa Barbara	SB	Sacramento Valley	SV
Santa Clara Valley	SCV	Pacific	PAC

7
Alaska	AK	Nevada	NV
Arizona	AZ	Oregon	OR
Eastern Washington	EWA	Utah	UT
Idaho	ID	Western Washington	WWA
Montana	MT	Wyoming	WY

8
Michigan	MI	West Virginia	WV
Ohio	OH		

9
Illinois	IL	Wisconsin	WI
Indiana	IN		

Ø
Colorado	CO	Missouri	MO
Iowa	IA	Nebraska	NE
Kansas	KS	North Dakota	ND
Minnesota	MN	South Dakota	SD

Canada
Maritime	MAR	Saskatchewan	SK
Newfoundland/Labrador	NL	Alberta	AB
Quebec	QC	British Columbia	BC
Ontario	ON	Northern Territories	NT
Manitoba	MB		

Non US stations should be logged as DX

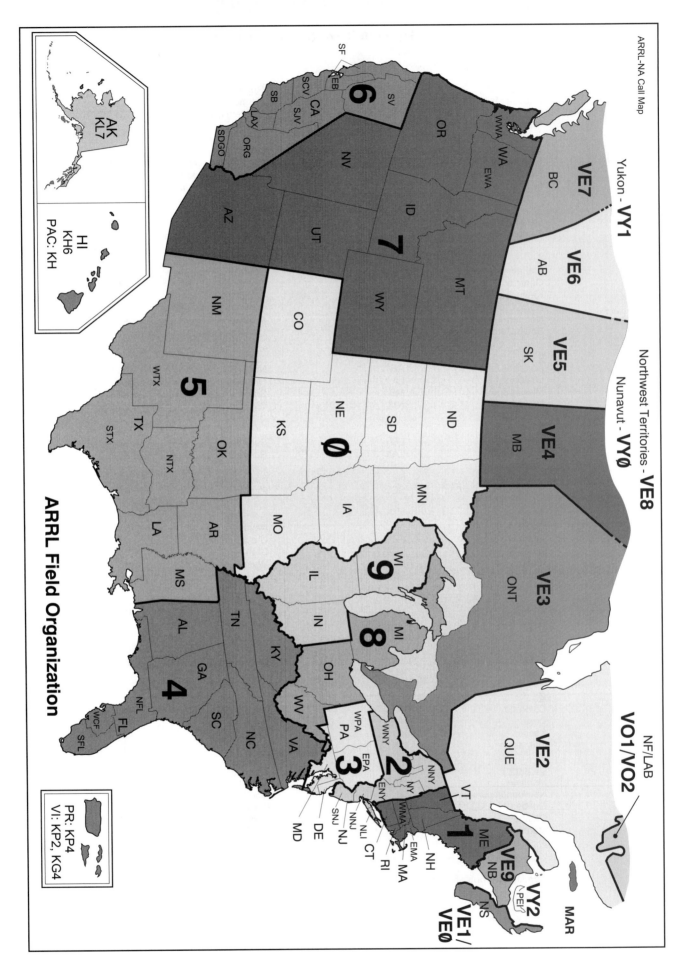

Morse Code Character Set[1]

A	didah	•—
B	dahdididit	—•••
C	dahdidahdit	—•—•
D	dahdidit	—••
E	dit	•
F	dididahdit	••—•
G	dahdahdit	——•
H	dididit	••••
I	didit	••
J	didahdahdah	•———
K	dahdidah	—•—
L	didahdidit	•—••
M	dahdah	——
N	dahdit	—•
O	dahdahdah	———
P	didahdahdit	•——•
Q	dahdahdidah	——•—
R	didahdit	•—•
S	dididit	•••
T	dah	—
U	dididah	••—
V	dididah	•••—
W	didahdah	•——
X	dahdididah	—••—
Y	dahdidahdah	—•——
Z	dahdahdidit	——••
1	didahdahdahdah	•————
2	dididahdahdah	••———
3	didididahdah	•••——
4	dididididah	••••—
5	dididididit	•••••
6	dahdididit	—••••
7	dahdahdididit	——•••
8	dahdahdahdidit	———••
9	dahdahdahdahdit	————•
0	dahdahdahdahdah	—————

At [@]	didahdahdidahdit	•——•—•	$\overline{AC}$
Period [.]:	didahdidahdidah	•—•—•—	$\overline{AAA}$
Comma [,]:	dahdahdididahdah	——••——	$\overline{MIM}$
Question mark or request for repetition [?]:	dididahdahdidit	••——••	$\overline{IMI}$
Error:	dididididididit	••••••••	$\overline{HH}$
Hyphen or dash [—]:	dahdidididah	—••••—	$\overline{DU}$
Double dash [=]	dahdidididah	—•••—	$\overline{BT}$
Colon [:]:	dahdahdahdididit	———•••	$\overline{OS}$
Semicolon [;]:	dahdidahdidahdit	—•—•—•	$\overline{KR}$
Left parenthesis [(]:	dahdidahdahdit	—•——•	$\overline{KN}$
Right parenthesis [)]:	dahdidahdahddidah	—•——•—	$\overline{KK}$
Fraction bar [/]:	dahdididahdit	—••—•	$\overline{DN}$
Quotation marks ["]:	didahdididahdit	•—••—•	$\overline{AF}$
Dollar sign [$]:	didididahdididah	•••—••—	$\overline{SX}$
Apostrophe [']:	didahdahdahdahdit	•————•	$\overline{WG}$
Paragraph [¶]:	didahdidahdidit	•—•—••	$\overline{AL}$
Underline [_]:	dididahdahdidah	••——•—	$\overline{IQ}$
Starting signal:	dahdidahdidah	—•—•—	$\overline{KA}$
Wait:	didahdididit	•—•••	$\overline{AS}$
End of message or cross [+]:	didahdidahdit	•—•—•	$\overline{AR}$
Invitation to transmit [K]:	dahdidah	—•—	K
End of work:	didididahdidah	•••—•—	$\overline{SK}$
Understood:	didididahdit	•••—•	$\overline{SN}$

Notes:

1. Not all Morse characters shown are used in FCC code tests. License applicants are responsible for knowing, and may be tested on, the 26 letters, the numerals 0 to 9, the period, the comma, the question mark, $\overline{AR}$, $\overline{SK}$, $\overline{BT}$ and fraction bar [$\overline{DN}$].

2. The following letters are used in certain European languages which use the Latin alphabet:

Ä, Ą	didahdidah	•—•—	
Á, Å, À, Â	didahdahdidah	•——•—	
Ç, Ć	dahdidahdidit	—•—••	
É, È, Ę	didahdidit	••—••	
È	didahdididah	•—••—	
Ê		—••—•	
Ö, Ô, Ó	dahdahdahdit	———•	
Ñ	dahdahdidahdah	——•——	
Ü	dididahdah	••——	
Ź	dahdahdidit	——••	
Z	dahdahdidididah	——••—	
CH, Ş	dahdahdahdah	————	

3. Special Esperanto characters:

Ĉ	dahdidahdidit	—•—••	
Ŝ	didididahdit	•••—•	
Ĵ	didahdahdahdit	•———•	
Ĥ	dahdidahdahdit	—•——•	
Ŭ	dididahdah	••——	
Ĝ	dahdahdidahdit	——•—•	

4. Signals used in other radio services:

Interrogatory	dididahdidah	••—•—	$\overline{INT}$
Emergency silence	didididahdah	••••——	$\overline{HM}$
Executive follows	dididahdidah	••—••—	$\overline{IX}$
Break-in signal	dahdahdahdahdah	—————	$\overline{TTTTT}$
Emergency signal	didididahdahdahdididit	•••———•••	$\overline{SOS}$
Relay of distress	dahdididahdididahdidit	—••—••—••	$\overline{DDD}$

Antenna Projects
from the pages of *QST*

By L. B. Cebik, W4RNL

Having a Field Day with the Moxon Rectangle

Good gain and a high front-to-back ratio are a couple of reasons to consider this antenna for Field Day use.

Field Day antenna installations tend to pass through phases. Phase 1 is the starter for any group: Get some antennas—usually dipoles and inverted Vs—into the air and see how well they perform. Phase 2 rests on an evaluation of the initial results. It generally consists of mechanical improvements to place the same or similar antennas higher using stronger materials. It also includes making better use of potential antenna supports at the site.

Real antenna design work usually begins with Phase 3. Based on the improved results with Phase 2 changes, the group begins to think about where they want the signals to go and how to get them there. At this stage, the group takes its first steps toward designing wire beams for the HF bands. (In Phase 4, we find the use of portable crank-up towers, rotators and multiband arrays. I'll not delve into Phase 4 in this article.)

Wire beams and arrays have one significant limitation: We can't rotate them. Therefore, we must resort to carefully planned aiming during installation. Still, we can only cover so much of the area across the country with the beamwidth available from gain arrays. Dreamers will always wonder if they could have garnered a few more contacts lost to the deep front-to-side ratio offered by most two-element Yagi designs.

So let's explore an alternative to the two-element wire Yagi, one that is only about 70% as wide, side to side, and which offers some other benefits as well: the wire Moxon Rectangle.

The Moxon Rectangle

In its most fully developed monoband form, a Moxon Rectangle outline looks like the sketch in Figure 1. A is the side-to-side length of the parallel driver and reflector wires. B is the length of the driver tails, while D is the length of the reflector tails. C is the distance between the tips of the two sets of tails. If any dimension of the Moxon Rectangle is critical, it is C. E, the total front-to-back length of the array, is simply the sum of B, C, and D.

The history of the Moxon Rectangle is itself fascinating.[1] Basically, it derives from early experiments with a square shape by Fred Caton, VK2ABQ, although the very first

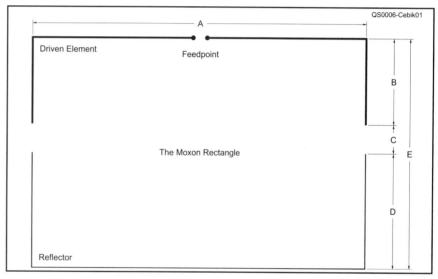

Figure 1—Outline of a Moxon Rectangle with various dimensions labeled. See the text for an explanation of the labels.

Table 1
Dimensions of Wire Moxon Rectangles for 80-10 Meters
All dimensions refer to designations in Figure 1. Dimensions are in feet and apply to #14 AWG bare-wire antennas.

Band	Frequency (MHz)	A	B	C	D	E
80	3.6	99.98	15.47	2.16	18.33	36.96
75	3.9	92.28	14.28	2.00	16.92	33.20
40	7.09*	50.69	7.82	1.15	9.35	18.32
20	14.175	25.30	3.87	0.62	4.70	9.19
15	21.225	16.88	2.56	0.44	3.14	6.14
10	28.3*	12.65	1.90	0.35	2.36	4.61

*Because of bandwidth versus wire-size considerations, 40- and 10-meter design frequencies are below the mid-band points to obtain less than 2:1 50- SWR over as much of the band as possible. See the text for alternative strategies.

experiments were performed in the 1930s. Les Moxon, G6XN, outlined in his classic *HF Antennas for All Locations*, a rectangular variant in which he remotely tuned the driver and the reflector.[2] Curious about the basic properties of the rectangle, I modeled and built variations of the design for about eight years, using wire and aluminum tubing.[3]

The Moxon Rectangle has three properties that recommend it for Field Day use:

• It is not as wide as an equivalent wire Yagi, because the two elements fold toward each other.

• It offers—with the right dimensions—a 50-Ω feedpoint impedance so no matching system is required (although use of a choke to suppress common-mode currents is always desirable).

• It presents a very useful Field Day pattern, with good gain and a very high F/B.

Figure 2 overlays the pattern for a typical two-element Yagi (reflector-driver design) and the Moxon Rectangle. The pattern may appear odd since it uses a linear decibel scale (rather than the usual log decibel scale) to enhance the detail at the pattern center. Although the Yagi has slightly more gain, the Moxon's deficit won't be noticeable in operation. Most apparent is the F/B advantage that accrues to the Moxon. In practical terms, the Moxon effectively squelches QRM to the rear. Of equal importance is the broader beamwidth of the Moxon. The azimuth pattern does not show deep nulls off the ends of the beam elements. Instead, the deep nulls are about 15 to 20° farther back. Signals off the beam sides are stronger than those of a Yagi, even though the rear quadrants themselves are that much quieter than the Yagi. (At low heights, from $^{3}/_{8}$ λ to 1λ, the Moxon's side gain ranges from 2 to 6 dB greater than that of a similarly positioned two-element Yagi.) As a result, the Moxon provides useful signal strength from one side to the other—as if it had good peripheral vision.

A Moxon Rectangle aimed in the general direction of the greatest number of potential Field Day contacts will generally gather signals from a broader sector of the horizon than most other antennas—with the bonus of good QRM suppression from the rear. Stations located near one of the US borders may discover that a basic, fixed Moxon Rectangle is all they need. For those stations located inland and needing coverage in all directions, I'll have a solution a bit later. But first, let's design a Moxon Rectangle.

Designing a Moxon Rectangle

The objective in designing a Moxon Rectangle is to produce a set of dimensions for the wire diameter used that yields maximum F/B, maximum gain and a 50-Ω feedpoint impedance at the design frequency. For this

[1]Notes appear on page 25.

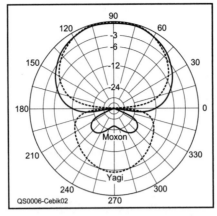

Figure 2—Relative free-space azimuth patterns at 14.175 MHz for a wire Yagi (driver and reflector) and a wire Moxon Rectangle. These patterns use a linear decibel scale to enhance detail at the pattern center (rather than the more usual log-decibel scale). Compare the pattern scale to that used in Figure 5.

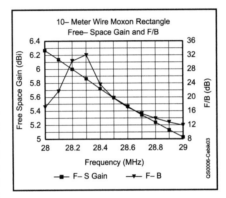

Figure 3—The pattern of free-space gain and 180° F/B across 10 meters for a #14 AWG wire Moxon Rectangle.

exercise, I chose #14 bare copper wire, perhaps the most popular Field Day antenna material. I also aligned the maximum F/B and 50-Ω resonant feedpoint frequencies. Of course, gain varies across the band as it does with any two-element parasitic array.

With these design criteria, Table 1 provides the dimensions of Moxon Rectangles for 80, 75, 40, 20, 15 and 10 meters—all potential Field Day bands of operation. The design frequencies are listed with the band of operation. Because the 40- and 10-meter bands are wide relative to the wire size used, I moved their design frequencies below the mid-band point in order to obtain low-end coverage at an SWR under 2:1.[4]

The Moxon Rectangle functions by virtue of the mutual coupling between parallel element segments and the coupling between the facing element tips. Hence, the gap between element tips (dimension C in Figure 1) is the most critical dimension. Measure the gap accurately and ensure that the spacing does not change over time. The other dimensions

follow from setting the gap in order to obtain the desired performance characteristics.

Figure 3 shows the gain and F/B curves for a 10-meter version of the #14 wire Moxon Rectangle, designed for 28.3 MHz. I chose 10 meters because even the first megahertz represents a very wide band. Note that the gain curve is nearly linear across the band. However, the F/B peaks near the design frequency and tapers off—more rapidly below the design frequency than above it. Figure 4 shows a similar curve for the 50-Ω SWR, with the rate of increase more rapid below the design frequency than above it.

There is no absolute need to align the maximum F/B frequency with the resonant 50-Ω feedpoint. We can move one or both of them by small adjustments in the antenna dimensions. To sample the rates of change in performance parameters relative to small changes in dimensions, I altered some dimensions of a 20-meter version of the antenna by one inch. (One inch at 20 meters is, of course, approximately equivalent to changes of four inches on 80, two inches on 40, and a half-inch on 10 meters.) In all cases, the gap (dimension C) is held constant.

• Decreasing or increasing the side-to-side dimension (A in Figure 1) raises or lowers the maximum F/B and the resonant feedpoint frequencies by about 40 kHz. For small changes in dimension A, the resonant feedpoint impedance does not change.

• Increasing or decreasing only the length of the driver tails (dimension B) by one inch lowers or raises the resonant frequency of the driver by about 70 kHz. The new resonant feedpoint impedance will be a few ohms lower (for an increase in driver length) than before the change. The frequency of maximum F/B will not change significantly.

• Increasing or decreasing only the length of the reflector tails (dimension D) by one inch lowers or raises the peak F/B frequency by about 70 kHz. The driver's resonant frequency will not significantly change, but the impedance will be higher (for an increase in reflector length) than before the change.

With these guidelines, you can tailor a basic Moxon Rectangle design to suit what you decide is best for your operation.

One of the realities of Field Day is that you will not operate your antenna in free space. Actual antenna heights over real ground may range from $^{1}/_{4}$λ to over 1λ, depending on the band and the available supports. To sample the operation of the Moxon Rectangle at various heights, I modeled a 10-meter version of the antenna at various heights, listed in Table 2 in terms of fractions of a wavelength. The performance of versions for other bands will not materially differ for equivalent heights.

Note that as the antenna height increases, the take-off angle (or the elevation angle of maximum radiation) decreases, as do the vertical and horizontal beamwidths between half-power points. These properties are in line with those of any horizontally polarized array. Hence, the gain increases slightly with antenna height increases. Figure 5 overlays the azimuth patterns for all of the heights in the table to demonstrate the small differences among them. Moreover, the feedpoint impedance of the antenna undergoes only small changes with changes in heights. Indeed, the excellent F/B performance at the low height of $^3/_8 \lambda$ holds promise for 40-meter and lower-frequency installations. The upshot of this exercise is that a Moxon Rectangle falls in the class of "well-behaved" antennas, requiring no finicky field adjustments once the basic design is set and tested.

Of course, you should always pretest your Field Day antennas using circumstances as close as possible to those you will encounter at the actual site. Testing over a prairie and operating in a forest can produce surprises (and problems) for almost any antenna. However, the semi-closed design configuration of the Moxon Rectangle tends to yield fewer interactions with surrounding structures than antennas with linear elements, an added advantage for Field Day operations.

A Direction-Switching Moxon Rectangle

If you live somewhere within the vast central region of the country, you may be interested in signals from both sides of the Moxon Rectangle. The antenna can accommodate you with fair ease. Following the design lead of Carrol Allen, AA2NN, we can design the Moxon Rectangle for direction-switching use.[5] Figure 6 shows the outline. Essentially, we create two resonant drivers using the same dimensions as for the basic antenna. Then we load the one we select as the reflector so that it becomes electrically long enough to perform as a reflector. Our loading technique employs a length of shorted 50-Ω cable. By bringing equal length stubs to a central point, we can switch them. The one we short becomes part of the reflector. The other one is connected to the main feed line and simply becomes part of the overall system feed line.

One switching caution: Use a double-pole double-throw switch so that you switch the center conductor and the braid of the coax lines used as stubs. When in use as a shorted stub, the line should not be electrically connected to the main feed line at all. A plastic box used to insulate the coax fittings from each other makes a good Field Day switch mount.

Table 3 lists the suggested dimensions for Field Day directional-switching Moxon Rectangles for 80 through 10 meters. Because two drivers are used, with their shorter tails, the overall front-to-back dimension (E) of each antenna is smaller than that of its one-way versions. The shorter front-to-back dimension lowers the feedpoint impedance by 5 to 7 Ω into the mid-40-Ω range, still a very good match for a coax feed line.

Table 3 also lists two stub lengths. The shorter one is the basic length of a shorted

Table 2
Relative Performance of a Wire Moxon Rectangle at Different Heights above Ground

Height (λ)	TO angle (Degrees)	Gain (dBi)	F/B (dB)	VBW (Degrees)	HBW (Degrees)	Feedpoint Z (R±jX Ohms)
Free-space	—	5.9	37.1	—	78	53 + j2
0.375	34	9.5	30.1	44	86	53 + j8
0.5	26	10.5	21.3	32	82	59 + j3
0.75	18	11.0	23.5	20	79	50 + j1
1.0	14	11.3	30.4	5	79	56 + j3

The modeled antenna is a 10-meter #14 AWG wire Moxon Rectangle at 28.5 MHz. Take-off (TO) angle refers to the elevation angle of maximum radiation. The 180° F/B is used in this table. Vertical bandwidth (VBW) and horizontal bandwidth (HBW) refer to the beamwidth between points at which power is down –3 dB relative to the maximum power. The feedpoint impedance (Z) is given in conventional resistance/reactance terms. See Figure 5 for comparative azimuth patterns.

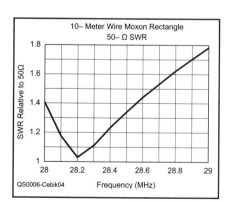

Figure 4—50-Ω SWR pattern across 10 meters for the #14 AWG wire Moxon Rectangle in Figure 3.

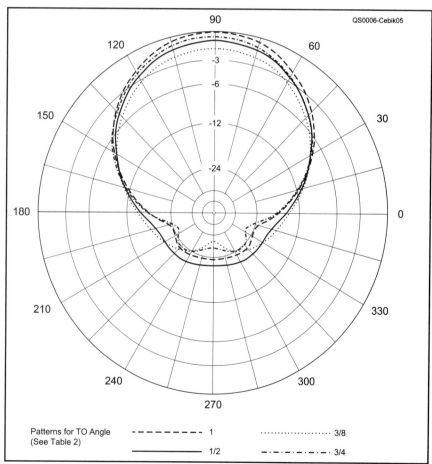

Figure 5—Typical azimuth patterns of a wire Moxon Rectangle at different heights (in wavelengths) above ground. Each azimuth pattern is taken at the elevation angle of maximum radiation (take-off angle).

50-Ω stub to achieve the required reflector loading. All of the designs required just about 65 Ω inductive reactance to electrically lengthen the reflector so that the maximum F/B frequency aligns with the driver resonant point. Hence, the basic stub length for the shorted stub is about 52.4°. Because you have a choice of cables with solid and foam dielectrics, you must multiply the listed length by the *actual velocity factor* of your stub cable. In general, solid-dielectric 50-Ω cables have velocity factors of 0.66 to 0.67, while foam cables tend toward a velocity factor of about 0.78. However, I have found significant departures from the listed values, so measuring the velocity factor of your line is a good practice. Otherwise, expect to cut and try lengths until you hit the right one.

Because the shorter length of the stub for some bands may leave them hanging high in the air, I have also listed the lengths of stubs that add a $1/2\lambda$ of line to them. The loading effect will be the same as for the shorter stub, but the lines may now reach a more convenient level for switching, especially in field conditions. It is wise to keep the lines suspended in the air, with the switch box hanging from a tree limb or tied to a post or stump. Again, multiply the listed values of longer lines by the velocity factor of the line you are actually using. Finally, be aware that coax stubs are not lossless and thus may slightly alter the performance of the array relative to the perfect lines used in models. In most cases, the differences will not be noticeable in practice.

The principles of reflector loading apply not only to Moxon Rectangles, but as well to wire Yagis, deltas, quads and a host of other two-element parasitic arrays. With good preplanning, they yield antennas simple enough to be manageable in the field. At the same time, you gain the benefits of a directional pattern that may nearly double your score. In non-scoring terms, a directional-switching array means more effective communication under almost all conditions.

Field Construction of a Moxon Rectangle

Despite their simplicity and low cost, wire beams can be ungainly. Hence, you should survey the Field Day site in advance—and if possible, practice raising and lowering the antennas. For the Moxon Rectangle, look for or plan for suitable supports to stretch the antenna at its corners. Of course, the higher the support, the better. Because the Moxon Rectangle is only about 70% the side-to-side width of a comparable two-element Yagi, its space requirements are relatively modest, allowing the site designer somewhat greater flexibility.

Figure 7 outlines two types of systems for supporting the Moxon Rectangle. Consider them to be only the barest starting points for a real system. The four-post system at the left is suitable for any band. The posts can be trees, guyed masts, or building corners. The rope terminating at the post can be tied off there, if the ring point is accessible. Or, run the rope over a limb or through an eyebolt so that the corner can be easily raised and lowered.

The ring at the end of the corner rope through which the wire passes is used to reduce mutual abrasion of the wire and rope and can be a simple loop in the rope or even a plastic bottleneck. Because the shape of the Moxon Rectangle is important, the corner bends should be locked. A short piece of wire that runs from main wire to tail, but which goes around the corner ring, can effectively keep the corner in place. A permanent installation might call for soldering the ends of the locking wire to the antenna elements, but a short-term field installation can usually do well with just a few twists of the locking wire on the element.

The two-post construction method is more apt to the upper HF bands. It uses a long pole, PVC tube, or similar nonmetallic structure to anchor the corner ropes. The corner rope can be terminated at the pole or passed through it and run to the post. The sketch shows a two-anchor mounting for the pole. The upper support ropes align the pole horizontally. Thus, the rope should be locked to the ring or other support to keep everything horizontal. Alternatively, you can brace the pole directly to the support post, tree, or mast so that it remains horizontal. The remaining attachment mechanisms are the same as for the four-post method of support.

The rope that separates the driver and reflector tails should not stretch. Its job is to maintain the tail gap spacing as securely as possible. In addition, since the degree of coupling between tails is a function of the wire diameter, the wire fold-back used to make an attachment loop in the element tails should be as tight and flat as possible without weakening the antenna wire. For added strain relief and dimensional precision for upper HF versions of the Moxon Rectangle, it is possible to place the nonmetallic pole at or inside the perimeter of the antenna. With some judicious use of electrical tape where the elements end along the pole, you can omit the tail-to-tail rope altogether. For a larger, lower, HF-band

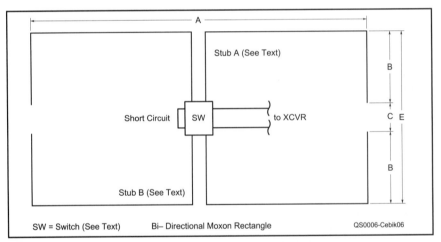

Figure 6—Outline of a direction-switching Moxon Rectangle, using transmission-line stub loading to electrically lengthen the reflector. See the text for details of the switching arrangement.

Table 3
Dimensions of Direction-Switching Wire Moxon Rectangles for 80-10 Meters

All dimensions refer to designations in Figure 6. Dimensions are in feet and apply to #14 AWG bare-wire antennas.

Band	Frequency (MHz)	A	B	C	E	Simple	Stub +1/2λ
80	3.6	99.98	15.47	2.16	33.10	39.78	176.39
75	3.9	92.28	14.28	2.00	30.56	36.72	162.82
40	7.09	50.69	7.82	1.15	16.79	20.20	89.56
20	14.175	25.30	3.87	0.62	8.36	10.10	44.80
15	21.225	16.88	2.56	0.44	5.56	6.75	29.92
10	28.3	12.65	1.90	0.35	4.15	5.06	22.44

Stub lengths are based on an inductively reactive load of 65 for the reflector element at the design frequency. Listed stub lengths are for 50- cable with a 1.0 velocity factor. Multiply listed lengths by the *actual velocity factor of the line* to obtain the final length.

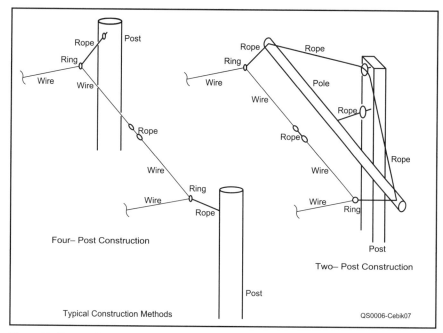

Figure 7—Four-pole and two-pole mounting arrangements for a wire Moxon Rectangle, shown only in barest outlines.

version of the antenna, you can use a rope that runs from each front ring to the corresponding rear ring and tape the driver and reflector tails wires to it.

For field use, lightweight coax (ie, RG-8X for 50-Ω applications) helps reduce the stress on the driven element(s) at the feedpoint. However, where conditions permit, supporting the element centers is advisable. In fact, slightly Ving the elements will normally produce no adverse effects in performance. However, if you contemplate a shallow inverted-V form of the antenna, pretest the assembly to assure that everything will work as planned.

Field Day antenna construction is a primary exercise in adapting easily obtainable materials to particular site configurations. Hence, it is not possible to provide universal guidance for every situation. However, these notes should get you started. Survey your local Home Depot and other such outlets for fixtures and nonmetallic connectors that might prove useful for a Field Day antenna. You may find them anywhere in the store. The plumbing and electrical departments are good starting places to find adaptable PVC fittings.

The Moxon Rectangle offers good potential for Phase 3 antenna improvements in Field Day installations. It is certainly not the only good antenna for this important exercise. The final decision you make in selecting an antenna should be the result of extended planning activities that review: (A) What is possible at the site; (B) what is possible with the available construction crew and (C) which antennas when properly oriented will improve communications the most from a given site. What you learn about various antennas that may be candidates for the next Field Day will serve you well in the long run—both at home and in the field.

Of course, the Moxon Rectangle—when it has done its Field Day service—need not be retired to storage awaiting next year's duty: It can serve very well in many home-station installations. The size and the signal pattern may be perfectly suited to the needs of at least some operators.

Notes

[1] For a more complete history, see L. B. Cebik, W4RNL, "Modeling and Understanding Small Beams: Part 2: VK2ABQ Squares and The Modified Moxon Rectangle," *Communications Quarterly*, (Spring, 1995), pp 55-70. There are a number of notes on this antenna type at my Web site (**http://www.cebik.com**) in the "Tales and Technicals" collection. As an example of a VHF version of the antenna, see Lee Lumpkin, KB8WEV, and Bob Cerrito, WA1FXT, "A Compact Two-Element, 2-Meter Beam," *QST*, Jan 2000, pp 60-63. Other VHF applications have appeared in *antenneX* an on-line magazine (**http://www.antennex.com**).

[2] Les A. Moxon, G6XN, *HF Antennas for All Locations* (RSGB, 1982), pp 67, 168, 172-175. Available from the ARRL, order no. 4300, $15. See the ARRL Bookcase in this issue for ordering information.

[3] For aluminum versions of the antenna, see L. B. Cebik, "An Aluminum Moxon Rectangle for 10 Meters," *The ARRL Antenna Compendium*, Vol 6 (ARRL, 1999), pp 10-13 and Morrison Hoyle, VK3BCY, "The Moxon Rectangle," *Radio and Communications* (Australia), Jul 1999, pp 52-53.

[4] If you want to use other wire sizes (including center-supported versions made from aluminum tubing in diameters up to well over an inch in diameter), a small GW Basic program is available that will ease the design work. The program's output is accurate to within under 0.5% relative to the *NEC-4* models used to derive the algorithms. You can download this program and explanatory text from **http://www.arrl.org/files/qst-binaries/** as *MOXONBAS.ZIP*.

A full account of the technique used to derive the program will appear in a forthcoming issue of *antenneX* (**http://www.antennex.com**). The program will also be added to the *HAMCALC* suite of GW BASIC electronics utility programs available from George Murphy, VE3ERP. Those having access to *NEC-Win Plus*, a *NEC-2* antenna modeling software package available from Nittany Scientific, can simplify the process of deriving dimensions and checking the resultant model. The model-by-equation facility of the spreadsheet input system permitted me to transfer the design equations directly into a model, which the user can set for any desired design frequency. The output will include both the dimensions and a standard *NEC-2* calculation of the antenna pattern and source impedance, with options for changing any of variables, including the wire conductivity, size, etc. A copy of the MOXGENE8 .NWP file is available among the examples at the NEC-Win Web site (**http://www.nittany-scientific.com**).

[5] Carrol Allen, AA2NN, "Two-Element 40-Meter Switched Beam," *The ARRL Antenna Compendium*, Vol 6 (ARRL, 1999), pp 23-25. See especially Carrol's improved method of stub construction.

*An ARRL Life Member and educational advisor, L. B. Cebik, W4RNL, recently retired from The University of Tennessee, Knoxville, to pursue his interests in antenna research and education, much of which appears at his Web site (**http://www.cebik.com**). A ham for over 45 years, his articles have appeared in several League publications including QST, QEX, NCJ and The ARRL Antenna Compendium. You can contact L. B. at 1434 High Mesa Dr, Knoxville, TN 37938-4443;* **cebik@utk.edu**.

Arthur S. Gillespie, Jr, K4TP

An Easily Constructed 30- and 40-Meter Trap Dipole Antenna

Spend a little time with some PVC and wire, and you'll be able to cover two bands with your own homemade dipole!

Discussions of trap-antenna construction and use have appeared in ARRL and other Amateur Radio publications for years. Using readily available materials, you can simply and inexpensively construct a 30- and 40-meter dipole antenna by assembling a pair of 30-meter traps. Most of the trap material you need is in stock at hardware and building-supply stores. For my station, I erected the dipole as an inverted **V**.

Trap Construction

See Figures 1 and 2 for the antenna dimensions and trap details, respectively. The traps use 1.5-inch Schedule 40 PVC pipe and mating end caps. Standard PVC cement glues the end caps to the pipe ends. The capacitors I used are Centralab 850S-50Z.[1] Because the power limit for the 30-meter band is 200 W PEP, lower-voltage ceramic capacitors of the proper value could probably be used. The value of a disc-ceramic capacitor designed for bypassing applications may vary considerably from its marked value, so if you plan to substitute capacitors, try 1-kV (or greater) capacitors after verifying their capacitance value. Use eyebolts and hardware made of corrosion-resistant metal such as stainless steel. Avoid using plated fittings; they won't last long without rusting. To allow the traps to breathe, drill a small hole in one end of each trap. This helps prevent moisture buildup within the trap.

After assembly, tune the traps to 10.125 MHz, the center of the 30-meter band. I did this by placing each trap—without antenna attached—on top of an empty cardboard box and tuning the trap to resonance using a dip meter. (If you have an antenna analyzer, use it—*Ed.*) First, check the calibration of the dip meter using a receiver. Make minor adjustment of the trap resonance by moving the coil turns closer together or farther apart. Once the traps are tuned properly, fix the windings in place with a light coat of PVC cement. When the cement is dry, check the trap resonance again. To hold the windings in place and lengthen trap life, cover the windings with an ultraviolet resistant sealant such as 3M Strip-Caulk (08578) available at auto parts suppliers, or use RadioShack sealing tape for outdoor connections (RS 278-1675). The sealant does not noticeably change the trap resonance.

Performance

[1]Suitable capacitors are available from RF Parts, 435 South Pacific St, San Marcos, CA 92069; **rfp@rfparts.com**; **www.rfparts.com**: 5-kV, 50-pF units (P/N 580050-5, factory part # HT50T500) are $12.95 each; 7.5-kV capacitors are $14.95 each. Another source is Surplus Sales of Nebraska, 1502 Jones St, Omaha, NE 68102; tel 800-244-4567, 402-346-4750, fax 402-346-2939; **grinnell@surplussales.com**, **www.surplussales.com**. (See page 101 of their catalog #8.—*Ed.*) Often these "doorknob" capacitors can be purchased inexpensively at hamfests and flea markets. High-voltage disc-ceramic capacitors are available from Mouser Electronics, 958 N Main St, Mansfield, TX 76063-4827; tel 800-346-6873, 817-483-4422, fax 817-483-0931; **sales@mouser.com**, **http://www.mouser.com** and Digi-Key Corp, 701 Brooks Ave S, Thief River Falls, MN 56701-0677; tel 800-344-4539, 218-681-6674, fax 218-681-3380; **http://www.digikey.com**.

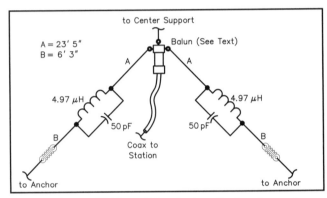

Figure 1—Inverted V 30- and 40-meter trap dipole antenna at K4TP.

I use a 1:1 balun at the center of the inverted **V**; the balun also serves as a center insulator. As shown, my antenna displays an SWR of 1.5 or less across the CW portion of the 40-meter band rising only to 2.8 at 7.3 MHz. On 30 meters, the SWR is 1.5 or less. My antenna has been in use for over two years and performs outstandingly on 30 and 40 meters. The traps should work equally as well in a standard dipole configuration; however, some minor changes in overall antenna length might be required to maximize the performance.

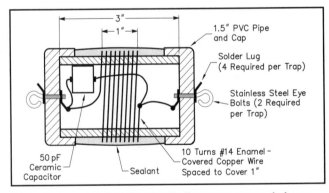

Figure 2—Trap construction details. Two traps are needed, one for each leg of the antenna. To ensure long life, use stainless-steel hardware and cover the winding with an ultraviolet-resistant sealer.

**618 Hillcrest Ave
Gastonia, NC 28052
artnfaye@bellsouth.net**

A Simple HF-Portable Antenna

By Phil Salas, AD5X

Tired of dragging that bulky old antenna tuner along on your vacation jaunts? Spare your suitcase and your pocketbook because this simple multiband wire antenna will get you on the air in a jiffy—with no extra gear required.

Every summer my wife (N5UPT), my daughter (AC5NF) and I spend about a week on Mustang Island off the coast of Corpus Christi, Texas. I always enjoy operating HF-portable when on vacation, and because Mustang Island is also known as IOTA NA092 (Islands On The Air, North American island number 92), getting on the air is even more fun! In case you're imagining typical DXpedition fare, you should know right from the start that we don't exactly rough it on Mustang Island. In fact, we always stay in a condo, which I request to be "the highest one available."

My first portable rig was a Kenwood TS-50, followed by an MFJ-9420 (see May 1999 *QST*). Last year I went deluxe and upgraded to an ICOM IC-706MKII. That little rig works dc to light—all bands and all modes, with goodies to boot. It it is an excellent choice for almost any type of portable operation.

I've experimented with several types of antennas on these outings—including Hamstick mobile whips, resonant dipoles and random-length wire dipoles fed through a tuner. I prefer resonant antennas so I don't have to worry about transporting and storing an antenna tuner. Of course, multiple dipoles or a handful of Hamsticks can take up a lot of room.

Last summer I used the multiband dipole described here with excellent results. If you're interested in a simple multiband wire that's easy to build and pack away, give this antenna a try.

The basic antenna covers all bands from 20-10 meters. You could increase its coverage, but the dimensions of a typical condo balcony seem to limit the lower frequency to 20 meters or so. If your operating site is larger, feel free to scale the antenna appropriately.

Basically, the antenna started as a full-size 20-meter dipole. I then inserted small in-line insulators to allow for multiband operation as shown in Figure 1.

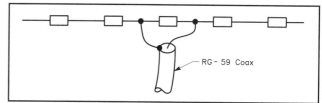

Figure 1—The concept began with a full-size dipole antenna that I "broke up" with small insulators.

The insulators are $3/8$-inch (diameter) by one-inch nylon spacers that can be found at most hardware stores. Each spacer is used as a "band switch" by drilling a small hole in each end and threading a short length of #14 bare wire (house wire) through each end, and attaching a short piece of wire terminated in an alligator clip. The clip, shown in Figure 2, is available at RadioShack stores (ask for part number 270-380).

I used #24 insulated wire for the dipole elements because it's lightweight and flexible. Obviously, any type of wire is fine. Use whatever you have on hand. The best way to determine the various segment lengths is to calculate the individual dipole lengths using:

L (feet) = 468/freq (MHz)

Tack solder the wire sections to the insulators, attach a feed line (RG-59 coax will do) and hang the dipole in a convenient place where it's easy to work on and adjust. Although the SWR meter method will work, to adjust the multiband dipole properly, beg, borrow or buy an antenna analyzer.

The entire antenna can be collapsed to a size that fits in the palm of your hand!

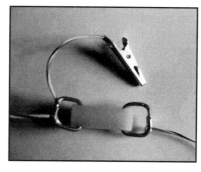

Figure 2—The band switches are constructed from nylon spacers, some wire and an alligator clip.

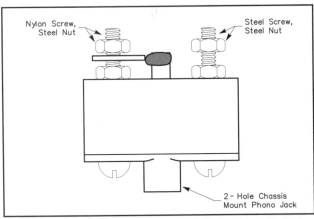

A photograph of my version of the center insulator.

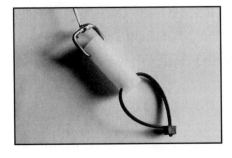

My design for the end insulator.

Figure 3 — I used an extra nylon spacer for the center insulator. I drilled the ends and attached a chassis-mount phono jack as shown. The nylon screw is used on one side to make sure that the phono jack's center conductor doesn't short to ground. I soldered #4 spade lugs to the inside ends of the 10-meter dipole elements so the dipole can be easily attached (and detached) to the center insulator. Feel free to use other center insulator designs as desired.

First, "unclip" all of the alligator clips and adjust the inner wire segments for the lowest SWR on your favorite part of the 10-meter band. The wires should be a bit long, so unsolder them on one end and trim them as follows:

New length = Original length × Measured low-SWR Frequency/Desired low-SWR frequency

Next, clip (attach) the inner pair of alligator clips and adjust the next segment for resonance on 12 meters using the formula and steps described previously. Continue this procedure for 15, 17 and 20 meters.

I know—you're adjusting your antenna low to the ground and your particular portable mounting location will undoubtedly vary. For our purposes it really doesn't matter. Most modern rigs can put out full power into a 2:1 SWR, so reasonable location-based SWR variations probably won't affect your rig's operation. If the SWR is really high, something's drastically wrong or you have the alligator clips set up for operation on the wrong band, etc. Incidentally, you can use a balun if you want to. I normally don't worry about feed line transformers when operating portable.

The antenna leg lengths I wound up with are shown below:
10 meters: 8 feet 3 inches on each side
12-10 meters: 10 inches on each side
15-12 meters: 1 foot 4 inches on each side
17-15 meters: 1 foot 8 inches on each side
20-17 meters: 3 feet 9 inches on each side

Each side is a total of 15 feet, 10 inches, for a total of 31 feet, 8 inches for the entire antenna.

Finally, if you want to electrically "shorten" your antenna, make the clip lead wires a little longer and wrap the excess wire around the insulators to make loading coils.

I used an extra nylon spacer for the center insulator. I drilled the ends and attached a chassis-mount phono jack as shown in Figure 3. The nylon screw is used on one side to make sure that the phono jack's center conductor doesn't short to ground. I soldered #4 spade lugs to the inside ends of the 10-meter dipole elements so the dipole can be easily attached (and detached) to the center insulator. Feel free to use other center insulator designs as desired.

Conclusion

If you need a simple portable antenna, spend an hour or two assembling this one. It's simple, cheap and a good performer. Simply adjust the clip leads for the desired frequency band and you're on the air—no tuner required! Sure, you have to make a quick trip to the balcony (or whatever) to change bands…but this is a vacation-oriented design, after all!

**1517 Creekside Dr
Richardson, TX 75081
ad5x@arrl.net**

By Robert Johns, W3JIP

A Ground-Coupled Portable Antenna

As the saying goes, "Imitation is the sincerest form of flattery." Here's a home-brewed antenna that proves the point.

This homebrew portable antenna for 40 through 6 meters is patterned after the ground-coupled design pioneered by Alpha Delta Commun-ications, Inc.[1] Instead of using radials, this antenna employs a simple and very small grounding system that needs no tuning.

The antenna described here is a quarter-wave vertical sitting on a tripod base. The vertical mast and the tripod are each made of 2-foot-long telescoping sections of $3/4$- and $5/8$-inch-diameter aluminum tubing.[2] The mast itself resonates on 10 meters; lightweight aluminum tubing sections are added to the top of the mast to tune the antenna to 12, 15 and 17 meters.[3] These added tubing lengths can be installed vertically or horizontally. The antenna is fed at the top of the tripod, making the base a part of the radiating system. A bungee cord stretched from the top of the tripod to a stake in the ground keeps the structure stable.

Beneath the foot of each tripod leg is a grounding strip $2^1/_2$ inches wide and about $3^1/_2$ feet long, made of aluminum tape.[4] These strips are simply laid on the ground and form one plate of a capacitor coupling RF from the antenna to the ground. That's the whole grounding system! When I read about this in *QST* (see Note 1), I was skeptical, but intrigued. The arrangement is similar to that of a mobile antenna system in which the car body acts as one plate of a capacitor coupling RF to the road and ground. This grounding system works: The antenna radiates well and the SWR is reasonably low on all bands. (The tripod and grounding strips can also be used with any vertical element or mobile whip you have.) A loading coil added between the aluminum tubing mast and the flattop permits operation on 20, 30 and 40 meters. With the coil positioned this far up the antenna, the entire 10 feet of tripod and mast are unloaded radiators on all HF bands.

[1]Notes appear on page 33.

Building the Tripod

The top of the tripod, Figures 1 and 2, makes it easy to set up. The three $5/8$-inch-diameter × 0.058-inch-wall aluminum tubes extending from the $1^1/_2$-inch PVC cap are permanently attached to it. To assemble the tripod, the legs slide over these tubes. A 3-inch-long, $3/8$-inch carriage bolt passes through a hole in the top of the PVC cap to support the vertical element. This bolt also grips the 4-inch-long aluminum tubes inside the cap to form the three sloping legs of the tripod. See Figure 2 and its caption for details on how to make this top cap.

A 50-Ω coaxial feed line attaches to the antenna via an SO-239 chassis connector

Figure 1—The top of the tripod with the bottom section of the mast connected to it. A bolt holds the three leg supports in the PVC cap slots. This bolt also passes through the $1^1/_2 \times 1^1/_2$-inch aluminum-angle piece that supports an SO-239 chassis connector for feed-line connection. A $1/_4$-inch hole in the cap top accepts the bungee-cord hook.

mounted on an aluminum angle bracket at the top of the PVC cap (see Figure 1). Make a $5/8$-inch-diameter hole in the bracket to accept the coax connector body; you'll also need to drill four small holes for the connector's mounting hardware. The $3/8$-inch bolt through the top of the cap keeps the aluminum bracket in place.

To assemble the tripod top, invert the cap so that you are looking down at the open end. Insert the carriage bolt through the $3/8$-inch hole in the cap, through the mounting hole in the aluminum angle and add a lock washer and nut to the bolt. Initially, thread the nut about an inch onto the bolt so that the bolt is still loose and its head is out of the cap. Insert the three aluminum tubes into the slots in the wall of the cap and down against the carriage bolt where it passes through the hole in the cap. Tighten the nut so that the carriage-bolt head squeezes the tubes outward and into the slots. Once the nut is hand tight, wriggle each tube to seat it snugly with its tip into the countersunk hole with the bolt. Tighten the nut until the round wall of the cap is slightly deformed into a triangular shape.

Each tripod leg consists of a 0.058-inch-wall, $3/4$-inch-diameter tube and a 0.058-inch-wall, $5/8$-inch-diameter tube that fits inside the $3/4$-inch tube. Each tube is two feet long; three can be made from 6-foot tubing lengths. Dimple each $3/4$-inch tube about one inch from each end. The dimple acts as a stop and prevents the smaller tube from penetrating any farther. Form the dimples using a couple of firm hammer taps on a center punch placed against the tube. When joining the tubes, push a bit when inserting the smaller tube so that the side of the dimple holds the smaller tube in place.

Ground Strips

Although mating two strips of aluminum tape with their sticky sides together might seem like a routine job, it's probably the most difficult part of building this antenna! The adhesive is quite sticky and unforgiving, and handling the long strips can be messy. Get an assistant to help you with this task. You'll need three strips.

See Figure 3. Cut a 7-foot length of tape from the roll and lay it down, sticky side up, on the floor or a large table. Have your helper press a piece of heavy (#12) solid wire or a thin dowel across the width of the strip at the $3^1/_2$-foot midpoint and hold it in place. Pick up one end of the strip and carry it over the midpoint, keeping it tight so that it doesn't sag and touch the lower half. Keep both ends of the strip aligned while your helper at the midpoint presses the top piece of tape against the lower, working their way toward you. Trim (or remove) the excess wire or rod and the ground strip is done. Don't worry if the strips aren't aligned perfectly.

The Mast

The 8-foot mast is made from two telescoping $3/_4$- and two $5/_8$-diameter × 0.058-inch-wall aluminum-tubing sections. Slot the ends of the $3/_4$-inch tubes so that they can be tightened around the smaller tubes with hose clamps.[5] To insulate the bottom $3/_4$-inch section from the $3/_8$-inch bolt in the tripod that sup-ports the mast, its lower end is equipped with a plastic insulator. As shown in Figure 4, the insulator is a 2-inch length of acrylic tubing. The lower end of the acrylic tube extends about a quarter inch below the aluminum tube and is slotted so that the mast can be tightened around the bolt. Drill a $5/_{32}$-inch hole through the upper end of this insulator and the aluminum tube to pass a #6-32 bolt and nut to hold the insulator in place.

After mounting the SO-239 coax connector on the aluminum angle strip, solder a 2-inch length of #14 bare solid copper wire to the connector's center terminal and bend it close to the $3/_8$-inch bolt in the tripod top. Then bend the wire up and parallel to the bolt and about a quarter inch from it. When the bottom section of the mast is placed over the bolt, place this wire between the aluminum tube and a hose clamp. As you tighten the clamp, it makes the electrical connection from the coax to the mast and squeezes the slotted aluminum tube and insulator tightly against the bolt.

With the mast on the tripod, an easy way to make frequency adjustments is to separate the mast from its bottom section and lower it to the ground. You can then reach the flattop and coil without tilting the mast. For this reason, I don't tighten this joint. I place a #6-32 bolt through the $5/_8$-inch tube which is the second section of the mast, about one inch from its lower end so that it doesn't slide very far in. I still use a hose clamp over the $3/_4$-inch tube, adjusting it to make a snug sliding fit for the upper mast.

For the top antenna sections, I use $5/_8$-inch-diameter thin-walled aluminum tubing used for aluminum clothes poles. This material is lighter and cheaper than the 0.058-inch-wall tubing used for the tripod and mast, but is strong enough. Short tubing sections can be joined together using 2-inch-long sleeves made from the $3/_4$-inch-diameter × 0.058-inch-wall aluminum tubing. You need two 2-foot, two $1^1/_2$-foot and two 1-foot lengths of the $5/_8$-inch thin-walled tubing, three couplings and a **T** joint to connect the flattop to the mast or the top of the coil. See Figure 5.

Bungee Tie-Down

The antenna is quite light, and even with the wide base of the tripod it needs to be stabilized against wind gusts or someone tripping over the coax feed line. A bungee

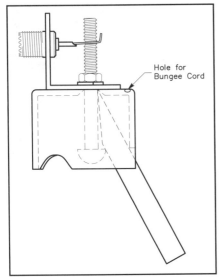

Figure 2—The tripod top cap. The three $5/_8$-inch-diameter aluminum tubes are 4 inches long, cut at a 30° angle at the end within the cap. From inside the cap, countersink a $3/_8$-inch hole in the cap top. This forms a trap that holds the ends of the aluminum tubes. Although only one of these leg supports is shown, all three are held between the bolt head, the three slots (either carved or filed in the wall of the cap) and the countersunk hole in the cap top. The slots in the cap are about $3/_8$-inch deep and wide enough to receive the aluminum tubes. An easy way to lay out the slots is to use the fluted handle from an outdoor water faucet as a template. The handle fits nicely against the cap and has six flutes about the circumference allowing you to mark three equally spaced locations.

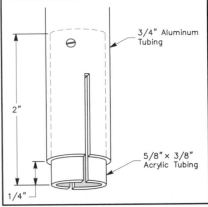

Figure 4—An acrylic (Plexiglas) tube insulates the antenna mast from the $3/_8 \times 3$-inch bolt that supports it. The tube has a $3/_8$-inch ID and $5/_8$-inch OD so that it slips over the supporting bolt and telescopes inside the lower $3/_4$-inch mast section. Both the aluminum tube and the insulator tube are slotted using a hacksaw so they can be tightened around the bolt with a hose clamp. To mount a mobile antenna on the tripod, cut a 2-inch length of 1-inch-diameter acrylic rod and drill and tap one end to accept the $3/_8 \times 16$ coarse-thread bolt of the tripod and $3/_8 \times 24$ fine threads at the other end for the base of a mobile whip.

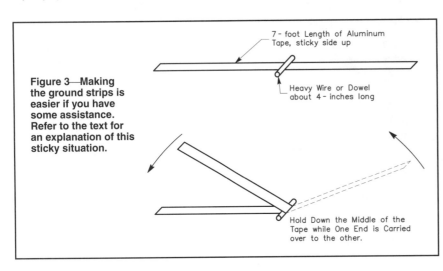

Figure 3—Making the ground strips is easier if you have some assistance. Refer to the text for an explanation of this sticky situation.

cord and a ground stake do an excellent job. The top of the tripod is about 3 feet high, so a ½- to ⅜-inch-diameter, 24-inch-long bungee cord works well. Any tent stake will do; drive it into the ground at an angle so it doesn't pull out easily. A special stake shaped like a large screw is ideal for this application.[6] It threads into the ground by hand and has a very low profile. (I leave the stake in the ground and my lawn mower doesn't even come close to striking it.) The stake won't go into hard, baked soil, however. For stability in such locales, or on pavement, hang some bricks, a rock or a jug of water beneath the tripod on the bungee cord.

Antenna Operation on 10 through 17 Meters

For 10-meter operation, set up the tripod and place one end of a ground strip under each tripod foot. The ground strips may be laid in any direction. Adjust the mast to a length of about 7.4 feet. This length is quite a bit less than a quarter wavelength and I believe it's because of its closeness to the ground and the thickness of the tripod. No top hat is used on 10 meters. Adjust the mast length to resonate the antenna at your desired 10-meter frequency.

Table 1 provides lengths for the thin-wall tubes that you add to the antenna, either as a flattop or a vertical, for operation on 12, 15 or 17 meters. No change to the ground system is needed when changing bands. Table 1 assumes that you will leave the mast set for 10-meter operation. This simplifies band changing, such as moving from 10 meters to 15 meters and returning to 10 meters. These changes are quickly made by just adding the tubing lengths for 15 meters and removing them to return to 10 meters—no measurements, no tools.

6-Meter Antenna Operation

For 6-meter operation, the tripod must be insulated from ground and the mast reduced to a length of 52 inches from tripod to tip; see Figure 6. No ground-coupling strips are needed. Simple insulators can be made from ½-inch CPVC pipe and couplings. Cut three lengths of pipe about 4 to 6 inches long and hammer each into a coupling. Cementing them isn't necessary; they will be a tight fit. The other side of the coupling fits well over the ⅝-inch-diameter aluminum-tubing leg. Adding these insulators to the tripod resonates the antenna in the 6-meter band with good SWR. You can change the operating frequency by adjusting the length of the mast only—you don't need to adjust the size of the tripod.

Building the Loading Coil

For operation on the 20, 30 and 40-meter bands, a loading coil must be added to the antenna. A large tapped coil is shown in Figure 7; it tunes the antenna to 20, 30, or 40 meters and permits you to tune to the higher-frequency bands without changing the lengths of the top hat. The coil has 13 turns of #8 aluminum wire wound on a 4-inch styrene pipe coupling.[7,8] This coil form is secured to a 7-inch-long, ½-inch-diameter CPVC pipe using 1¼-inch-long, #6-32 brass or stainless steel machine screws and nuts. I like to reinforce the ½-inch pipe by hammering a 2-inch length of ½-inch wood dowel into each end. This allows me to tighten the nuts and bolts without flattening the pipe. These bolts also secure the ends of the 13-turn coil. Using a marking pen, I made black marks on the coil to identify the fifth and tenth turns. The marks serve to locate the proper tap points without having to count coil turns each time.

Inside the styrene coil form is a ridge. Use a chisel or file to remove about a 1-inch-long section of this ridge to allow the CPVC pipe to lie flat against the inside of the form. Drill ⁷⁄₆₄-inch holes at the ends of the styrene coil form and through the ½-inch pipe, then bolt them together as shown in Figure 7. Take a 16-foot length of aluminum wire, bend a loop at one end of it, attach the loop to one of the bolts and wrap the form as neatly as possible with 13 turns of wire, without bends, spacing the turns to fill the form. Wrap the end of the 13th turn

Table 1
Length of thin-wall tubes needed for operation on 10 through 17 meters.

Band (Meters)	Length of Flattop (ft) and Number of Sections	Length of Vertical Top (ft)
10	0	0
12	1 × 2	1.5
15	2.5 × 2	3.5
17	3.5 × 2	5.5

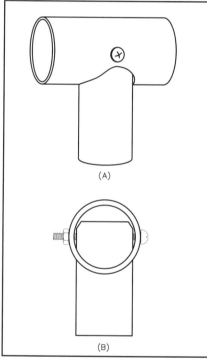

Figure 5—A T is needed to make the flattop. The ¾-inch-diameter horizontal tubing has a 0.058-inch wall and accepts the ⅝-inch-diameter thin-wall tubes. The ⅝-inch-diameter vertical piece has a 0.058-inch wall and fits into the ¾-inch tube at the top of the coil form. To make a ⅝-inch hole in the ¾-inch tube, drill a hole then expand it with a ⅝-inch-diameter or larger countersink. (This process is heavy work for a countersink, so use a little lubricating oil.) Before drilling a ⁷⁄₆₄-inch hole for a #6-32 bolt through the T, assemble the two pieces and squeeze them together tightly in a vise, making them perpendicular. To mount a flattop on the mast without the coil, first place a ¾-inch-diameter coupling sleeve over the ⅝-inch-diameter top of the mast and fit the T into that coupling.

Figure 6—Here, the antenna is set up for use on 6 meters. The tripod construction remains the same, using legs approximately 4 feet long, but the mast has been shortened. No ground strips are needed and the legs are insulated from ground by ½-inch CPVC pipe extensions at their feet.

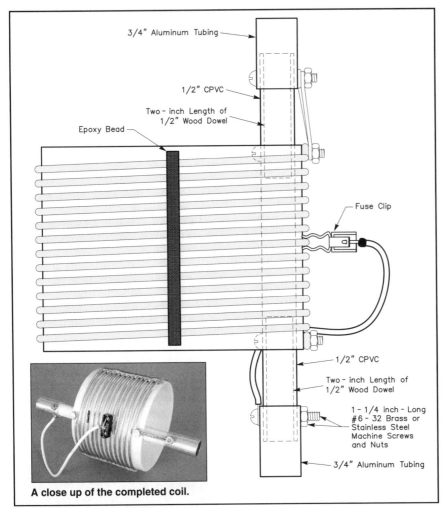
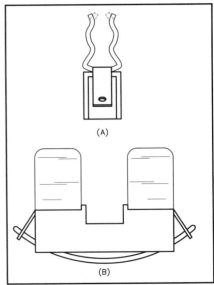

Figure 7—To make the loading coil, 13 turns of heavy aluminum wire are spaced to fill the form. Secure the coil ends using the same bolts that hold the plastic pipe inside the 4-inch styrene coil form. Mount the coil on the mast with a ³⁄₄-inch-diameter aluminum sleeve at the bottom of the plastic pipe; the tap wire is also connected here. An identical sleeve at the top of this plastic pipe connects to the thin tubing for the top vertical section, or to an aluminum T to hold the flattop.

Figure 8—Making the tap connection to the coil. At A, the ends of the jaws of a 5-mm cartridge-fuse holder are bent inward (dotted lines) to grip the heavy wire of the coil. A side view of the fuse holder is shown at B. Bend the solder lugs at the ends of the fuse holder to accept a wire passing through them and beneath the fuse-holder base. When this wire is in place, bend the lugs farther up against the ends of the holder and solder them. Strip a ¹⁄₄ inch of insulation from the tap wire and solder it to the wire joining the lugs beneath the fuse holder. Round off any sharp points or rough edges with a file, because you'll be gripping this connector tightly for attachment to and removal from the coil.

around the bolt at the other end of the coil form and cut off the excess wire.

To tighten the wire on the form, clamp the form in a vise, grab the coil turns between both hands and progressively rotate the coil from one end to the other several times. This makes the turns tight enough to stay in place as you even out their spacing. To hold the turns in place permanently, run three ribs of epoxy the length of the coil. Use metal/concrete epoxy which has black resin and white hardener, making a dark gray mix that is easy to see against the white background of the coil form. One of these ribs is visible in Figure 7. To make nice straight ribs, first place strips of tape on each side of an intended rib location, apply the epoxy and remove the tape before the epoxy hardens.

Several types of alligator-clips will fit between the coil turns without touching neighboring turns, but I prefer to use a tap connection made from a fuse holder; see Figure 8.[9] After bending the fuse-holder-jaw tips, bend the jaws themselves to make them fit the wire tightly, but remain easy to attach and remove. Suit yourself as to how tight a grip they should have. Join the tap connector to the sleeve at the bottom of the coil form using a 9-inch length of stranded, insulated #14 copper wire, with a solder lug at the end. Use a similar piece of wire to join the top of the coil to the sleeve at the top of the coil form.

The sleeve at the coil bottom joins the coil to the mast. It is a 1¹⁄₂-inch-long, ³⁄₄-inch-diameter, 0.058-inch-wall aluminum-tubing piece. Insert the bottom of the ¹⁄₂-inch CPVC pipe halfway into this sleeve and drill a ⁷⁄₆₄-inch hole through the sleeve and pipe. Fasten them together with a 1-inch-long, #6-32 brass or stainless steel machine screw and nut. The wire to the tap connector is attached with this same screw.

Antenna Operation with the Coil

To use the antenna on 40 through 10 meters, shorten the mast to 6 feet 2 inches and connect the coil to the mast. Atop the coil, add an element consisting of two horizontal 3¹⁄₂-foot lengths of ⁵⁄₈-inch-diameter tubing, or a single 7-foot vertical piece of tubing. With the full 13 turns of the coil, and part of an extra turn supplied by the tap wire, the antenna will likely resonate in the middle of the 40-meter band. To operate at the low end of the band, add a 1-foot length of tubing to one side of the flattop or the vertical tubing section. See Figure 9 for approximate dimensions of the assembled antenna.

It may seem as though Table 2 has some errors because it lists a greater number of coil turns for operation on 15, 12 and 10 meters than for 17 meters! You're right—something strange is going on. It's because there are *two resonant frequencies* for each setting of the coil tap. Figure 10 shows the two paths that RF can take in the antenna. The upper part of the coil and the top hat provide the lower frequencies; the lower half of the coil provides the higher frequencies. A coil this large has considerable capacitance to free space, so it's not just an end-loading inductor at the higher frequencies. The antenna bandwidth is good, the SWR low and the antenna performs well on these bands. The charm of this coil system is that you can change bands by just moving the

Table 2

This table identifies the number of coil turns (counted from the top of the coil) required to resonate the antenna on the 40- through 10-meter bands. These coil-tap settings are provided as a starting point only because installation conditions vary. To raise the antenna's operating frequency, reduce the number of turns used; to lower the operating frequency, increase the number of turns.

Band (Meters)	Number of Coil Turns
40	13
30	7.1
20	3.1
17	2
15	5
12	7
10	13

tap on the coil, without any adjustments to the mast length or the flattop. And a bonus: With 13 turns on the coil, the antenna works on 40 and 10 meters simultaneously.

The coil settings of Table 2 may need some minor adjustments if a vertical top section is used instead of the flattop. In general, the SWR is lower with the flattop and the antenna is easier to handle.

Power-Handling Capability and Safety

Because of the large coil and tubing used, you might be tempted to run high power with this antenna. I suggest you don't. The antenna may take it, but people can't. At high-power levels, dangerous RF voltages on the antenna are within range of physical contact. I have used the antenna at a 100-W level, but even that requires care and supervision.

Other Possibilities

With the tapped coil, this antenna can be tuned to any frequency from 7 to 40 MHz when operated on the ground-coupled tripod, and up to 110 MHz with the tripod insulated from ground.

The antenna also may be used with a longer mast for greater efficiency, or with a shorter mast when space is restricted. Even though the short version is only about 6 feet high, you can't use it indoors because it must be coupled to earth ground. The taller antenna gets out better, but band changing is more complicated. If operation on 75 and/or 80 meters is a must, you can add another coil to

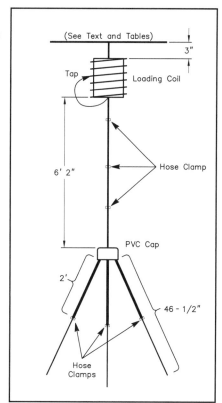

Figure 9—Approximate dimensions of the assembled antenna with the tripod, mast, loading coil and top hat.

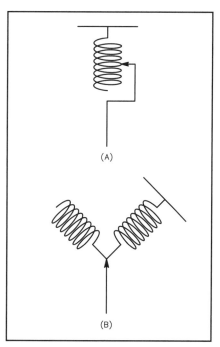

Figure 10—At A, the upper part of the antenna includes the coil, the adjustable tap and the top hat. The bottom of the coil is free and not connected to anything else. At B, this has been redrawn to show the two antenna circuits with the two resonant frequencies that are present. The upper half of the coil has a lower resonant frequency because of the length of the top hat above it.

the antenna just below the 40-meter coil and change antenna frequencies with the 40-meter tap. Adding a coil made of 20 close-wound turns of #12 enameled wire wound on a 4-inch styrene form similar to the one in Figure 7 will allow you to tune the antenna from about 3.5 to 3.8 MHz, and from about 3.8 to 4.1 MHz with the top hat reduced to one 3.5-foot section and one 2-foot section. Six ground-coupling strips will provide a lower SWR on 80. A small vertical like this is not very effective for short-skip ragchewing, however. A λ/4-wire draped over bushes, flower beds or low tree branches offers more high-angle radiation.

Notes

[1] Rick Lindquist, N1RL, and Steve Ford, WB8IMY, "Compact and Portable Antennas Roundup," 'Alpha Delta Outreach/Outpost System,' Product Review, QST, Mar 1998, pp 72-73.
[2] Twelve feet of each tubing size is needed. The aluminum tubing is available from Texas Towers and Metal and Cable Corp. See their ads elsewhere in this issue.
[3] The thin-walled $5/8$-inch-diameter aluminum tubing is available from Home Depot and hardware stores as aluminum clothes poles, each about seven feet long.
[4] Adhesive-backed aluminum tape $2^{1}/_{2}$ inches wide is available from Home Depot stores in the heating-vent section.
[5] You may want to consider using an antioxidant at the tubing joints. Antioxidant compounds available from electrical wholesale supply houses, Home Depot and hardware stores include Noalox (Ideal Industries Inc, Becker Pl, Sycamore, IL 60178; tel 800-435-0705, 815-895-5181, fax 800-533-4483) and OX-GARD (GB Electrical, 6101 N Baker Rd, Milwaukee, WI 53209; tel 800-558-4311). Use either sparingly; a thin coat is sufficient.—Ed.
[6] Aluminum angle $1^{1}/_{2} \times 1^{1}/_{2} \times 1/_{16}$-inch thick is available from hardware and Home Depot stores. The green plastic ground stake that threads into the ground has the name "Twizelpeg" stamped into it, and is available at camping supply stores.
[7] The #8 aluminum wire is RadioShack #15-035.
[8] The coupling is available from Home Depot in the drainage pipe section, and also from large plumbing or swimming pool distributors. The couplings are actually $4^{1}/_{2}$ inches in diameter and made from polystyrene, a very low-loss insulator.
[9] RadioShack #270-738.

Bob Johns, W3JIP, is an old gadgeteer who likes to play with antennas and coils. You can contact Bob at PO Box 662, Bryn Athyn, PA 19009; ksjohns@email.msn.com.

Photos by Joe Bottiglieri

By Joe Everhart, N2CX

The NJQRP Squirt

This reduced-size 80-meter antenna is designed for small building lots and portable use. It's a fine companion for the Warbler PSK31 transceiver.

At one time, 80 meters was one of the more highly populated amateur bands. Lately, it has become significantly less popular because much DXing has moved to the higher frequencies and many suburban lot sizes are too small to accommodate a full 130-foot, λ/2 antenna for the band. That's unfortunate, because 80 meters has lots of potential as a local-communication band—even at QRP levels. The recently published Warbler PSK31 transceiver can serve as a great facilitator for close-in QRP communication without much effort.[1] What's really needed to complement the Warbler for this purpose is an effective antenna that fits on a small suburban plot. Because PSK31 (which the Warbler uses) is reasonably effective even with weak signals, we can trade off some antenna efficiency for practicality.

What's a Ham to Do?

I investigated a number of antenna possibilities to come up with a practical solution. One intriguing candidate is the magnetic loop. Plenty of design information for this antenna is presented in *The ARRL Antenna Book* and at a number of Web sites.[2, 3] To obtain high efficiency, however, the loop must be 10 feet or more in diameter and built from $1/2$-inch or larger-diameter copper pipe. The loop needs a very low-loss tuning capacitor and a means of carefully tuning it because of its inherently narrow bandwidth. Another configuration, the DCTL, may be a solution, but it's likely not very efficient.[4]

An old standby antenna I considered is the random-length wire worked against ground. If it is at least λ/4 long (a Marconi antenna) or longer, it can be reasonably efficient. Shorter lengths are likely to be several S units down in performance and almost any length end-fed wire needs a significant ground system to be effective. Of course, you may not need much of a ground with a λ/2 end-fed wire, but it's as long as a center-fed dipole.

Vertical antennas don't occupy much ground space, but suffer the same low efficiency as the end-fed wire if they are practical in size.

Probably the easiest antenna to use with good, predictable performance is the horizontal center-fed dipole. Unfortunately, as mentioned earlier, the usual 80-meter λ/2 dipole is too large for many lots. But all is not lost! The dipole can be reduced to about a quarter wavelength without much sacrifice in operation (see the sidebar, "Trade-Offs"). Furthermore, if the dipole's center is elevated and the ends lowered—resulting in an inverted **V**—it takes up even less room. This article describes just such a dipole: the NJQRP Squirt.

V for Victory

You can think of the Squirt as a 40-meter, λ/2 inverted-**V** dipole being used on 80 meters. Figure 1 is an overall sketch of the antenna; Figure 2 is a photograph of a completed Squirt prior to erection. The Squirt has two legs about 34 feet long separated by 90° with a feed line running from the center. When installed, the center of the Squirt should be at least 20 feet high, with the dipole ends tied off no lower than seven feet above ground. This low antenna height emphasizes high-angle NVIS (Near Vertical-Incidence Skyware) propagation that's ideal for 80-meter contacts ranging from next door out to 150 or 200 miles. And that's where 80 meters shines! With the Squirt's center at 30 feet and its ends at seven feet, the antenna's ground footprint is only about 50 feet wide.

One nice feature of a λ/2 center-fed

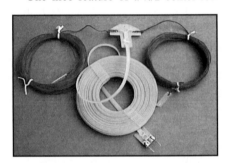

Figure 2—An assembled Squirt ready for installation.

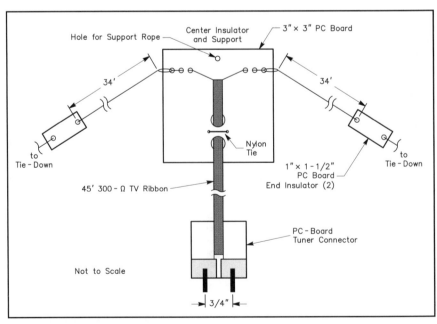

Figure 1—General construction of the 80-meter Squirt antenna.

[1]Notes appear on page 37.

dipole is that its center impedance is a good match for 50- or 75-Ω coax cable (and purists usually use a balun). Ah! But the Squirt is only λ/4 long on 80 meters, so it *isn't* resonant! Its feedpoint impedance is resistively low and reactively high. This means that feeding the antenna with coax cable would create a high SWR causing significant feed-line loss. To circumvent this, we can feed the antenna with a low-loss feed line and use an antenna tuner in the shack to match the antenna system to common 50-Ω coax cable. I'll have more to say about the tuner later.

I use 300-Ω TV flat ribbon line for the feed line. Although a better low-loss solution is to use open-wire line, that stuff is not as easy to bring into the house as is TV ribbon. Using TV ribbon sacrifices a little transmitted signal for increased convenience and availability. If you feel better using open-wire line, go for it!

Using Available Materials

It's always fun to see what you can do with junkbox stuff, and this antenna is one place to do it. See the "Parts List" for information on materials and sources.[5] For instance, the end and center insulators (see Figure 1) are made of $1/16$-inch-thick scraps of glass-epoxy PC board. For the antenna elements, I use #20 or #22 insulated hookup wire. Although this wire size isn't recommended for use with fixed antennas, I find it entirely adequate for my Squirt. Because it's installed as an inverted **V** antenna, the center insulator supports most of the antenna's weight making the light-gauge wire all that's needed. The small-diameter wire has survived quite well for several years at N2CX. This is not to say, of course, that something stronger like #14 or #12 electrical house wire couldn't serve as well.

The 300-Ω TV ribbon can be purchased at many outlets including RadioShack and local hardware stores. Once again, if you want to use heavier-duty feed line, do so. The only proviso is that you may then have to trim the feeder length to be within tuning range of the Squirt's antenna tuner.

The End Insulators

I used $1/2 \times 1 1/2$-inch pieces of $1/16$-inch PC board for the Squirt's two end insulators. As with everything else with the Squirt, these dimensions are not sacred; tailor them as you wish. If you use PC board for the end insulators, you have to remove the copper foil. This is easy to do once you've gotten the knack. Practice on some scraps before tackling the final product. The easiest way to remove the foil without etching it is to peel it off using a sharp hobby knife and needle-nose pliers. Carefully lift an edge of the foil at a corner of the board, grasp the foil with the pliers and slowly peel it off. You should become an expert at this in 10 or 15 minutes. Drill $1/8$-inch-diameter holes at each end of each insulator for the element wires and tie-downs.

Tuner Feed-Line Connector

The tuner end of the feed line is terminated in a special connector. Because the TV-ribbon conductors aren't strong, they'll eventually suffer wear and tear. This connector provides needed mechanical strength and a means of easily attaching the feed line to the tuner. In addition to some PC-board material, you'll need four or five inches of #18 to #12 solid, bare wire. Refer to Figure 3 and the accompanying photographs in Figures 4 and 5 for the following steps.

Take a $1 1/8 \times 1 3/4$-inch piece of single-sided PC board and score the foil

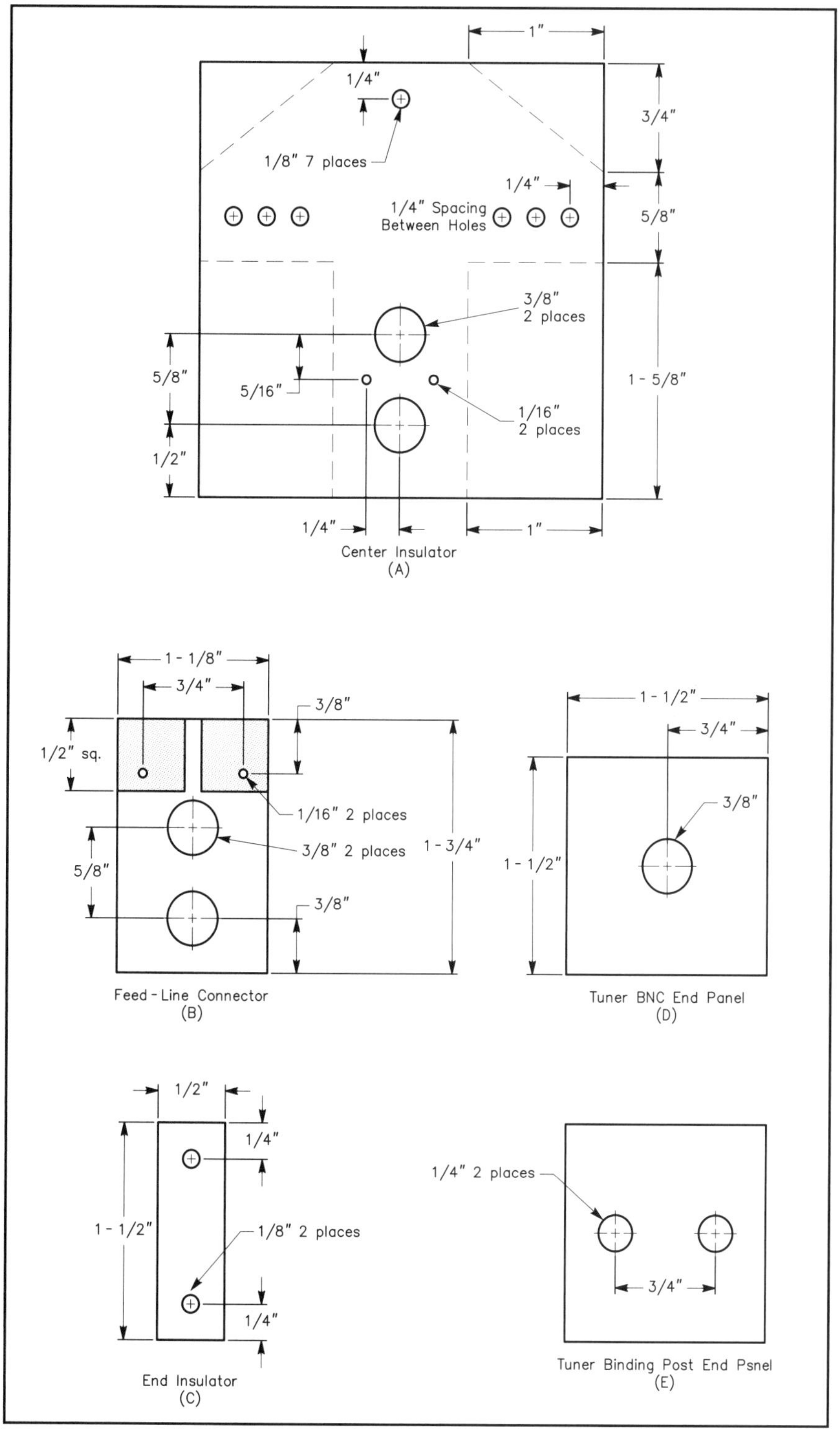

Figure 3—Hole sizes and locations for the various PC-board pieces. See Note 5.

35

Trade-Offs

One of the unfortunate consequences of shrinking an antenna's size is that its electrical efficiency is reduced as well. A full-size dipole is resonant with a feedpoint impedance that matches common low-impedance coax quite well. This means that most transmitter power reaches the antenna minus only 1 dB or so feed-line loss. However, when the antenna is shortened, it is no longer resonant. A *NEC-4* model for the Squirt shows that its center impedance on 80 meters is only about 10 resistive, but also about 1 k capacitive. This is a horrendous mismatch to 50- cable, and feed-line loss increases dramatically with high SWR. The Squirt uses 300- TV ribbon for the feed line with an inherently lower loss than coax. This loss is much less than if coax were used, but it's still appreciable. Calculated loss with 300- transmitting feed line is about 7.7 dB (loss figures are hard to come up with for receiving TV ribbon) so the feed line used doubtless has more than that.

Although this sounds discouraging, it's *not fatal*. You have to balance losing an S unit or so of signal against not operating at all! Consider that the Squirt, even with its reduced efficiency, is still better than most mobile antennas on 40 and 80 meters. So for local communication (a low-dipole's forte), using PSK31 and the Squirt is quite practical.

If you don't already have an antenna, the Squirt's a good choice to get your feet wet when using PSK 31. Once you get hooked, you'll probably want a better antenna. If you have the room, put up a full-size dipole; you'll see the improvement right away. If you can't do that, use a lower-loss feeder on the Squirt, such as good-quality open wire.—*Joe Everhart, N2CX*

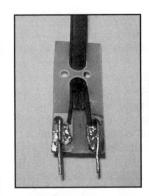

Figure 4—The pad side of the homemade feed-line-to-tuner connector.

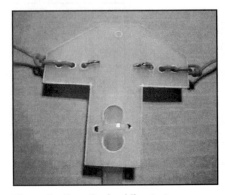

Figure 5—Here the feed-line-to-tuner connector is shown attached to the binding posts of the Squirt antenna tuner.

about $1/2$ inch from one end; remove the $1 1/4$-inch piece of foil. Now score the remaining foil so you can remove a $1/8$-inch-wide strip at the center of the board, leaving two rectangular pads as shown in Figures 3B and 4. Drill two $1/16$-inch-diameter holes in the copper pads spacing the holes about $3/4$-inch apart. Drill two $3/8$-inch holes at the connector midline about $5/8$-inch apart, center to center, to pass the feed line and secure it.

Cut two pieces of #18 to #12 wire each about three inches long. Pass one wire through one of the $1/16$-inch holes in the connector board and bend over about $1/4$-inch of wire on the nonfoil side. Solder the wire to the pad on the opposite side and cut the wire so that about one inch of it extends beyond the connector. Repeat this procedure with the second wire. Next, strip about two inches of webbing from between the feed-line conductors and loop the feed line through the two $3/8$-inch holes so that the free ends of the two conductors are on the copper-pad side. Strip each lead and solder each one to a pad. You now have a solid TV-ribbon connector that mates with the binding-post connections found on many antenna tuners. Figure 6 shows the connector mated with a Squirt tuner.

Center Insulator

Strip all the foil from this 3-inch-square piece of board. Use Figure 3A as a guide for the hole locations. The top support hole and the six wire-element holes are $1/8$-inch in diameter; space the wire-element holes $1/4$-inch apart. The feed-line-attachment holes are $3/8$-inch diameter spaced $1/2$-inch apart, center to center; the two holes alongside the feed-line-attachment holes are $1/16$-inch diameter. These $1/16$-inch holes accept a plastic tie to secure the feed line. I trimmed the insulator shown in Figure 2 from its original 3-inch-square shape to be more esthetic. Your artistic sense may dictate a different pattern.

Bevel all hole edges to minimize wire and feeder-insulation abrasion by the glass-epoxy material. You can do this by running a knife around each hole to remove any sharp edges.

Putting It All Together

The Squirt is simple to assemble. Once all the pieces have been fabricated, it should take no more than an hour or two to complete assembly. Begin with the center insulator. Cut each of the two element wires to a length of about 34 feet. Feed the end of one wire through the center insulator's outer hole on one side, then loop it back and twist around itself outside the insulator to secure it. Now loop it through the other two holes so that the inner end won't move from normal movement of the wire outside the insulator. Repeat the process for the other insulator/wire attachment. Separate several inches of the TV-ribbon feed-line conductors from the webbing; leave the insulation intact except for stripping about $1/2$ inch from the end of each wire. Pass the TV ribbon through both $3/8$-inch holes. Strip a $1/2$-inch length of insulation from each dipole element, then twist each feeder wire and element lead together and solder the joints. It might be prudent also to protect the joint with some non-contaminating RTV or other sealant. Finally, loop a nylon tie through the holes alongside the feeder and tighten the tie to hold the feeder securely. A close-up of the assembled center insulator is shown in Figure 6.

Attach the end insulators to the free ends of the dipole wires by passing the wires through the insulator holes and twisting the wire ends several times to secure them.

So that the antenna/feed-line system can be tuned with the Squirt tuner, the 300-Ω feed line needs to be about 45 feet long. If you use a different tuner, you may have to make the feed line longer or shorter to be within that tuner's impedance-adjustment range.

Tuner Assembly

This tuner (see Figures 7 and 8) is about as simple as you can get. It's a basic series-tuned resonant circuit link-coupled to a coaxial feed line. At C1, I use a 20 to 200-pF mica compression trimmer acquired at a hamfest (you *do* buy parts at hamfests, don't you?), although almost any small variable capacitor of this value should serve. The inductor, L1, consists of 50 turns of enameled wire wound on a T68-2 iron-core toroidal form. An air-wound coil would do as well, although it would be physically much larger. Figure 8 shows the tuner built on an open chassis made of PC board. My prototype uses several PC-board scraps: a 2×3-inch piece for the base plate, two $1 1/2$×$1 1/2$-inch pieces for each end plate (refer to Figure 3). A $1/2$-inch square piece of PC board (visible just beneath the capacitor in Figure 8) is glued to the base plate to serve as an insulated tie point for the connection

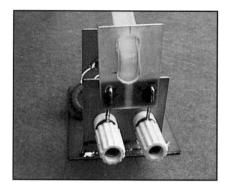

Figure 6—View of an assembled center insulator fashioned from a 3×3-inch piece of PC board from which all the foil has been removed.

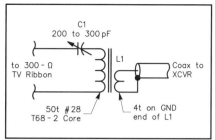

Figure 7—Schematic of the Squirt antenna tuner. See the accompanying Parts List.

Figure 8—This Squirt tuner prototype uses a 2×3-inch piece of PC board for the base plate, two 1^1/$_2$ × 1^1/$_2$-inch pieces for end plates and a ^{1}/$_2$-inch square piece as a tie point for the toroid and tuning capacitor.

between the toroid (L1) and tuning capacitor (C1). The tuner end plates are soldered to the base plate to hold a pair of five-way binding posts and a BNC connector at opposite ends. L1 and C1 float above electrical ground, connected to the TV ribbon. One end of L1's secondary (or link) is grounded at the base plate and the coax-cable shield. The hot end of L1's secondary winding is soldered to the coax-connector's center conductor.

Tuner Testing

C1 tunes sharply, so it's a good idea to check just how it tunes before you attach the tuner to an antenna. You can simulate the antenna by connecting a 10-Ω resistor across the binding posts. If you use an antenna analyzer as the signal source, a ^{1}/$_4$-W resistor such as the RadioShack 271-1301 is suitable. But if you use your QRP transmitter, you need a total resistance of 8 to 10 Ω that will dissipate your QRP rig's output, assuming here it's 5 W or less. Four RadioShack 271-151 resistors (two series-connected pairs of two parallel-connected resistors) provide a satisfactory load if you don't transmit for extended periods. Or, you can make up your own resistor arrangement to deliver the proper load. Adjust C1 with an insulated tuning tool to achieve an SWR below 1.5:1.

Once the tuner operation is verified using the dummy antenna, it's ready to connect to the Squirt. Tuning there will be similarly sharp, and a 2:1 SWR bandwidth of about 40 kHz or so can be expected as normal.

A Multiband Bonus

Although the Squirt was conceived with 80-meter operation in mind, it can double as a multiband antenna as well. The simple Squirt tuner is designed to match the antenna only on 80 meters. However, a good general-purpose balanced tuner such as an old Johnson Matchbox or one of the currently popular Z-match tuners (such as an Emtech ZM-2) will give good results with the Squirt on any HF band. The Squirt prototype was recently pressed into service at N2CX on 80, 40, 30, 20 and 15 meters for several months. It worked equally as well as a similar antenna fed with ladder line. Although no extensive comparative tests were done, the Squirt has delivered QRP CW contacts from coast to coast on 40, 20 and 15 meters and covers the East Coast during evening hours on 80 meters.

Build one! I'm sure you'll have fun building and using the Squirt!

Parts List

Squirt Antenna

Numbers in parentheses refer to vendors presented at the end of the list.

1—3×3-inch piece of ^{1}/$_{16}$-inch-thick glass-epoxy PC board for the center insulator (1)
2—^{1}/$_2$×1^1/$_2$-inch pieces of PC board for the end insulators (1)
1—1^7/$_8$×1^3/$_4$-inch piece of PC board for the feed-line connector (1)
2—34-foot lengths of #20 (or larger) insulated hookup wire (2)
1—6-inch length of #16 (or larger bare) copper wire; scrounge scraps from your local electrician.
1—45-foot length of 300- TV ribbon line (2)

Squirt Tuner

1—2×3-inch piece of PC board for base plate (1)
2—1^1/$_2$-inch-square pieces of PC board for end plates (1)
1—^{1}/$_2$-inch-square piece of PC board for the tie point (1)
1—200- to 300-pF (maximum) mica compression trimmer (3)
1—T68-2 toroid core (3)
2—Five-way binding posts (2)
1—55-inch length of #26 or 28 enameled wire (2 and 3)

Note: You can use ^{3}/$_{16}$-inch-thick clear Plexiglas for the Squirt's end and center insulators. Commonly used as a replacement for window glass, Plexiglas scraps can be obtained at low cost from hardware stores that repair windows.

Vendors

1. HSC Electronic Supply, 3500 Ryder St, Santa Clara, CA 95051; tel 408-732-1573, **www.halted.com**
2. Local RadioShack outlets or **www.RadioShack.com**
3. Dan's Small Parts and Kits, Box 3634, Missoula, MT 59806-3634; tel 406-258-2782; **www.fix.net/~jparker/dans.html**

Notes

[1] Dave Benson, NN1G, and George Heron, N2APB, "The Warbler—A Simple PSK31 Transceiver for 80 Meters," QST, Mar 2001, pp 37-41.
[2] R. Dean Straw, N6BV, The ARRL Antenna Book (Newington: ARRL, 1997, 18th ed), pp 5-9 to 5-11.
[3] **www.alphalink.com.au/~parkerp/nodec97.htm**; **www.home.global.co.za/~tdamatta/loops.html**
[4] **home.earthlink.net/~mwattcpa/antennas.html**
[5] Full-size templates are contained in SQUIRT.ZIP available from **www.arrl.org/files/qst-binaries/**.

You can contact Joe Everhart, N2CX, at 214 New Jersey Rd, Brooklawn, NJ 08030; **n2cx@arrl.net**.

Photos by the author

By Bob Clarke, N1RC

Gain without Pain—A Beam Antenna for Field Day

Stymied by lack of a tower or need an antenna for Field Day? By using low-cost PVC pipe, electrical wire, and coax you can build a variety of delta-loop-based beams that you can hang from a friendly tree. The cost? About $5 per dB![1]

Lacking space enough for a tower and possessing an over-active imagination, I've spent a fair amount of time over the last few years looking at various multielement directive arrays. I needed an antenna that could provide enough gain to be heard in DX contests, fit in my small back yard, meet the spouse's approval, and above all, be cheap.

Living here in New England, I only wanted gain in two directions: to the northeast to work Europe in DX contests and to the southeast to work the US in domestic contests; these two directions account for the bulk of the QSOs in these contests. For DX contests, I also wanted a decent front-to-back ratio (F/B) so I could hear Europeans while attenuating the loud signals that arrive from the Midwest as the HF bands open up to the rest of the US. Other directions could be filled in using my trap dipole.

Over the years, I've considered and rejected assorted ways of getting gain: pairs of phased verticals (too many radials), ZL specials (requires end supports in exactly the right directions), Bobtail curtains (again end supports) and phased Bobtail curtains (yet more end supports), and assorted others. I started looking at quads when K1DG gave me a vintage two-element tribander. Dissatisfied with the compromises in gain and F/B caused by using equal spacing for the 20, 15 and 10-meter bands, I started exploring ways of improving performance while keeping the spacing for all three bands the same. I came up with the idea of using phasing lines to turn a two-element quad into a two-element phased array using quad elements.

Now phased arrays[1] have some interest-

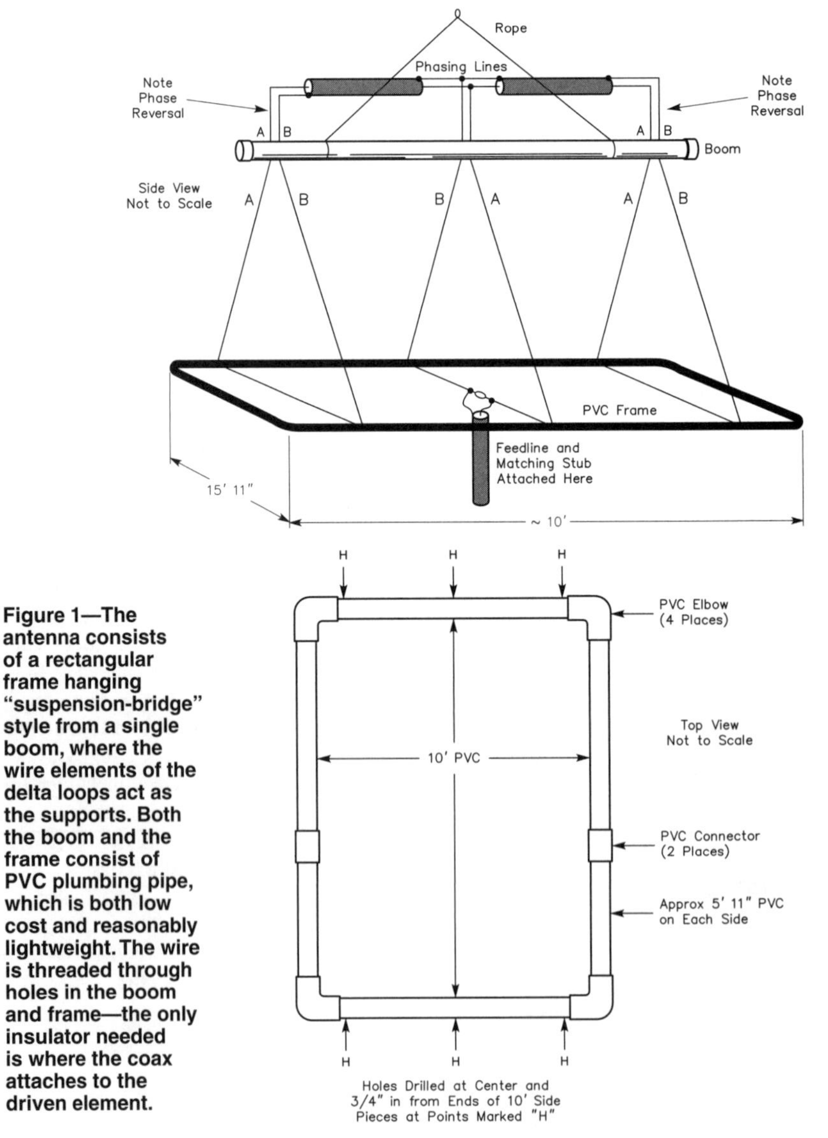

[1]Notes appear on page 41.

Figure 1—The antenna consists of a rectangular frame hanging "suspension-bridge" style from a single boom, where the wire elements of the delta loops act as the supports. Both the boom and the frame consist of PVC plumbing pipe, which is both low cost and reasonably lightweight. The wire is threaded through holes in the boom and frame—the only insulator needed is where the coax attaches to the driven element.

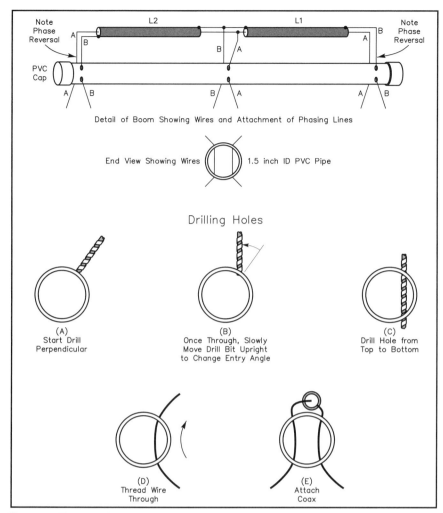

Figure 2—The two wires at the tops of the loops are threaded through pairs of holes in the boom. The bare ends are brought out at the top, where they connect to the two phasing lines. In the 3-element phased quads, the connections at the ends of phasing lines L1 and L2 are reversed at the outer elements to add an additional 180 degrees of phase

What is a Binomial Array?

A Binomial Array has a current distribution among its elements that follows the coefficients of a binomial expansion. The current ratios for two, three, and four element arrays can be generated by using Pascal's Triangle, where each entry is the sum of the two entries immediately above it:

```
            1
         1     1
      1     2     1
   1     3     3     1
```

From rows two, three and four, the current ratios in the elements of a 2-element array are 1-1; in a 3-element array, 1-2-1; and in a 4-element array, 1-3-3-1. An ideal binomial array has no sidelobes.

An example of a binomial array in Amateur Radio use is the Bobtail Curtain. In this antenna, all the current flows through the center element and half flows to each of the outside elements through the quarter-wave phasing lines at the top to provide a 1-2-1 ratio.

I took this technique and extended it to delta loops (and quads!) by feeding the center element at the bottom and using phasing lines at the top to route half the current to each of the outside loops. The phasing lines also introduce a phase shift that causes the three loops to act as an end-fire array, that is, unidirectional. The phase shifts between elements were optimized for F/B. This type of antenna is formally known as "An End-Fire Array of Delta Loops with Binomial Current Distribution."

The 1-2-1 binomial array can also be thought of as a pair of two-element arrays with the center elements superimposed. Through a principle of antenna design called pattern multiplication, this squares the pattern of two 2-element designs but uses only three elements. The overall pattern is the product of three patterns: element gain (loop gain over isotropic radiator) × (two element array pattern) × (two element array pattern).

ing pros and cons. Among the pros is that you can design one for reasonable gain and very high F/B with almost arbitrary spacing between elements. The con is that they can be narrow band, and the pattern can reverse when tuning from the low end to the high end of a wide band such as 15 meters. Generating the correct phase shifts can be tricky, since the transmission lines and the loads interact, and many articles have been written on this topic.[2]

To prove the concept of phased quads, I settled on phased delta loops, with the initial concept of using two equal-sized loops as a phased array hung from a single boom with a rectangular frame of PVC plumbing pipe spreading the bottom corners of the array. The boom could be hung from a tree branch by a single rope and rotated by a tag line or attached to a tower with a switchbox controlling the phasing lines to reverse the direction of the pattern.

The fact that you can arbitrarily set the spacing lets you design the antenna around available materials, in this case low-cost PVC pipe, which comes in 10-foot lengths. To simplify construction, I settled on a boom length conforming to the standard 10-foot length of pipe. As a CW operator, having to use a narrowband antenna wasn't a drawback.

My initial idea was to build a 2-element phased array but, as often happens in engineering, elegance crept in, this time on little EZNEC[3] feet (see the sidebar, "Delta Loop Arrays You Can Build"), and the two element concept morphed into a 3-element design with a binomial current distribution (see the sidebar, "What is a Binomial Array?").

As proof-of-concept antenna, I decided to build a 15-meter array with 3 elements on a 10-foot boom with the goal of using it the Fall 2000 CQWW CW DX Contest. Antenna jockeys will notice that 3 elements on a 10-foot boom is a bit cramped for 15 meters, and that 10 meters might have been a better choice for this boom length. This was a 3-way compromise: a 3-element design, 15 meters was the band that would provide the most payback in a DX contest, and a 10-foot boom to simplify construction. I did also simulate a version of this antenna on a 12-foot boom and found only a few tenths of a dB difference in gain—not worth the effort to extend the length of the boom and frame.

Many different antennas and antenna designs can be built using this method. Table A lists EZNEC analysis results and dimensions for several examples: the 3-element example described here; a pair of 10 meter arrays: one on a 10-foot boom and one on a 13-foot boom optimized for gain; and a pair of 2-element arrays: a 2-element 20-meter phased array on a 10-foot boom and a 2-element 15-meter parasitic array designed by N6BV. Note that there is no practical difference in the forward gain between the 3-element binomial array described in this article and N6BV's 2-element parasitic design; the parasitic design has a broader bandwidth but the binomial array has 10 dB better F/B.

Also note that Table A includes lengths and impedances for matching stubs, which attach to the driven element in parallel with the feed line. N6BV has found that the impedance of 450- open-wire line is actually 405 , so this value is used. The lengths of the phasing lines are critical—these are based on their impedance and velocity factor. The 53.5- phasing line is created with RG8X or equivalent coax with a foam dielectric. The 202.5- phasing line impedance for one of the 10-meter antennas is based on using two identical sections of 405- open-wire lines in parallel. The 150- phasing line impedance for the 20-meter antenna is based on using two 300- TV type twinlead sections in parallel.

The matching stub itself simply attaches in parallel with the feed line. You can substitute another transmission line for the stub provided that it has the correct input impedance. The shortest stubs are open-wire line or 300- twinlead due to their higher Z_0.

The wire used for the antennas is not critical provided that it can support the weight of the PVC frame. For the 15-meter antennas, a 50-foot roll of 3-conductor,

Analyzing the Delta Loop Array

The delta-loop arrays described here consist of two or three elements. Figure A shows the electrical configuration of the antenna. In Figure B you can see the *EZNEC* analysis results on various bands. Table A provides the analysis results in tabular format.

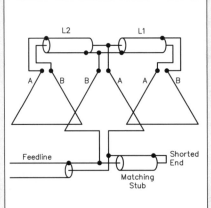

Figure A—The 3-element array of delta loops feeds all the current through the center element and then routes half to each of the outer elements via transmissions lines connected from the top of the center element to the tops of the outer elements to generate the 1-2-1 current distribution. These transmission lines also provide the necessary phase shifts to create the unidirectional or end-fire pattern.

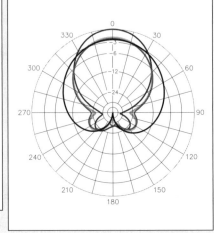

Figure B—*EZNEC* analysis results. The 2-element, 15-meter parasitic array pattern is shown in red. The black plot represents a 2-element, 20-meter phased quad on a 10-foot boom. The violet plot shows a 3-element 10-meter phased quad on a 10-foot boom. The green pattern is the product of a 3-element, 15-meter phased quad on a 10-foot boom.

Table A
EZNEC Analysis Results

Configuration	Frequency (MHz)	Forward Gain (dBi)	F/B Ratio (dB)	Loop Length (feet)	Element Spacing (feet)	Phasing Line L1 (feet)	Phasing Line L2 (feet)	Phasing Line Z_0 (Ω)	Matching Stub (feet)	Matching Stub Z_0 (Ω)
3 element phased 15-meter quad on 10-foot boom	21.050	6.86	39.94	47.76	4.95	10.25	23.25	53.5	9	405
3 element phased 10-meter quad on 10-foot boom	28.150	7.12	36.87	35.73	4.95	7.73	17.31	53.5	2	300
3 element phased 10-meter quad on 13-foot boom	28.050	8.85	30.77	35.82	6.25	15.9	18.9	202.5	2.2	405
2 element phased 20-meter quad on 10-foot boom	14.050	6.54	49.45	71.52	9.90	38.4	N/A	150	5.5	300
2 element parasitic 15-meter quad on 10-foot boom	21.050	6.77	29.34	47.76	7.00	N/A	N/A	N/A	2.3	405

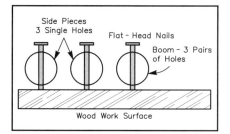

Figure 3—To align the holes in the PVC, first drill one hole in the end of each piece, about 1/2 inch from the end. Then insert a flat head nail though the holes and nail it about 1/2 inch into a piece of scrap wood to ensure than the nail and thus the hole are perpendicular to a level work surface. Do the same with the other three pieces. Now three pieces should be side by side nailed to the same piece of scrap wood. You can now drill the remaining holes and know that they will be parallel by drilling them perpendicular to your work surface.

14-gauge NMC (nonmetallic conductor) costs about $10 and provides the wires for all 3 elements for the 10- or 15-meter antennas—all you have to do is remove the outer sheath. The bare (ground) conductor can be used as the driven element since this requires the most soldered connections, and the insulated wires for the outer loops.

Finally, *EZNEC* lets you scale these antennas for different frequencies, such as 6 or 2 meters. But that's a future project!

Construction

Construction is simple: the antenna can be assembled and raised in a single afternoon by one person; no construction crew needed. What's more, it can be built so that it can be disassembled and moved, stored, or just used for contest weekends.

The antenna itself is built "suspension-bridge" style (Figure 1). A single PVC boom provides the top support and the elements are spread by a rectangular PVC frame that hangs from the boom, supported by the antenna wires. There are no insulators at the corners of the loops—the wire is threaded through holes drilled in the PVC. The only insulator is at the bottom of the center element where the feed line is connected.

As Figure 1 shows, you need one piece of PVC for the boom, two for the sides, and two each for each end, which are 15 feet 10 inches long for this 15-meter array. Each piece consists of one 10-foot section and a 5-foot, 10-inch section, joined and glued together by a PVC adapter. The corners consist of PVC elbows glued to the 10-foot sidepieces. When you assemble the frame, do not glue the end pieces. Instead, insert

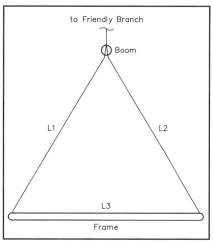

Figure 4—Hoist the entire antenna so that the frame is off the ground, but still low enough that you can comfortably work on it. Adjust the loops by moving the wires through the frame until the frame is level. Once the frame is level, the loops are equilateral triangles and your adjustment is done. Attach the feed line and matching stub and hoist the antenna into place. Enjoy!

them for a snug fit and then drill a 1/4-inch hole completely through the pipe and elbow. Make a cotter pin by inserting a 6-inch piece of scrap wire through the assembled pipe and elbow and secure it by twisting the ends of the wire together. Removing these wires will allow you to disassemble and roll up the antenna for transport (to the Field Day site, for example) or storage between contests.

For the case of the 3-element Binomial array, you need to drill three pairs of holes through the boom, in parallel, from bottom to top, one pair for the two wires at the top of each loop (Figure 2). You also need to drill a corresponding single hole in each of the sidepieces. The easiest way to do this is to lay all three pieces side by side with the ends aligned and measure and mark all the holes (Figure 3).

Once you drill one end of the three pieces, drive a nail through each into a piece of scrap wood. This will ensure that you have the holes at the middle and far ends in alignment (Figure 4). You can also glue the elbows on at this point, ensuring that they are perpendicular to the entry holes for the wires.

To assemble the antenna, cut three equal lengths of wire (lengths are in the sidebar, "Delta Loop Arrays You Can Build") to form the loops. Thread each wire through each of the two sides first and then bring the ends up through the holes in the boom. Attach the phasing lines by soldering and insulating, paying careful attention to the sense of the connections (in some designs, the phasing lines may have a twist). Coil the slack in the phasing lines and attach the coil to the boom with electrical tape.

Once the antenna is assembled, you'll have to adjust the position of the spreaders on the loops to ensure that the wire in each loop forms an equilateral triangle. This is probably the trickiest part of the job! Do this by hoisting the antenna in the air until the spreader clears the ground. You can then slide the wire through the holes in the PVC side supports as necessary to ensure that the spreader is level (Figure 4). Once the spreader is level, the loops will be equilateral triangles. (Since the width of the frame is the length of one side, the only way for the frame to be level is for all three sides of the triangle to be the same length, that is, L1=L2=L3.) If the holes aren't too large, there should be sufficient tension due to the weight of the frame to keep the wires from sliding.

Performance

The performance of the prototype was delightful: European signals were several dB louder than on my trap dipole and QRM from US stations was negligible! I have the ability to raise and lower the antenna when not in use. Given the high winds at my location (about 1/2 mile from the ocean and a clear shot to the northeast), the ability to lower the antenna enhances its survivability! And if I want to use it at a different location, I only have to remove the wire "cotter pins" and roll it up for transport.

Notes
[1] My main reference in developing this design was "Chapter 4, Arrays of Point Sources" in J. D. Kraus, *Antennas*, McGraw-Hill, 1988, which discusses pattern multiplication, phased arrays, and binomial current distributions.
[2] See, for example, Roy Lewallen, "The Simplest Phased Array Feed System... That Works", *ARRL Antenna Compendium*, Volume 2
[3] *EZNEC 3.0* Antenna Design Software by Roy Lewallen is a wonderful toy. As N6BV commented, "How else can you build and debug an antenna in mid-winter without going outdoors?"

Bob Clarke was first licensed in 1973 as WN1RLI and has a BSEE from the Massachusetts Institute of Technology. In addition to experimenting with antennas, he collects and operates boatanchors. He currently works in new product marketing at Analog Devices, Inc. in Wilmington, MA where he is responsible for the definition of RF ICs. He can be reached at 301 Washington St, Gloucester, MA 01930-4815; **BClarke@alum.mit.edu**.

By David G. Byrd, KD7VA

The Arkansas Catfish Dipole

Fishin' for DX on your next ham radio outing? Whether Field Day or just for fun, this portable 20-meter antenna is easy to assemble and easy on your wallet. Go ahead—reel 'em in!

Two of my favorite pursuits are Amateur Radio and fried catfish. As a matter of fact, as I prepare this article, the odor of frying Arkansas Catfish envelops me from the nearby kitchen. That wonderful smell—and my favorite "fishing pole"— had a lot to do with the title of this article.

This past winter, I spent most of December and January in southern Arkansas helping to care for my wife's parents. Although we didn't have much warning before departure, I did pack my backup HF transceiver and 2-meter hand-held. I also took my MFJ-259B Antenna Analyzer so I could worry about the antennas after I had arrived and surveyed possible antenna sites.

Eight hours after we arrived, southern Arkansas was hit with the worst ice storm in more than 30 years. Because most of the tree limbs were broken and on the ground, all thoughts of hanging wire antennas from the remaining ice-burdened limbs went south (further south) for the winter. I decided to construct a 20-meter antenna that could be assembled indoors (during the bad weather) and erected later with minimal effort.

Necessity...

With no "ham store" in Magnolia, Arkansas, on-site hardware pickings were slim. The local Radio Shack stocked a few CB antennas and 50-foot lengths of RG-58 coax, but nothing ham-specific. The store normally carried 20-foot telescoping steel masts, but even these were out of stock. I would have to gather any remaining components from Ace Hardware and Wal-Mart.

I purchased the #20 enamel wire and the two 14-foot cane poles at Wal-Mart. The cane poles were varnished and separated into three five-foot (or less) sections for transport. The remaining parts were purchased at Ace Hardware but should be available at most hardware stores. There are many mast choices with widely varying lengths and costs, but after the ice storm had wiped out hundreds of TV antennas the previous week, I had to get creative and use a telescoping pool cleaning pole.

Assembly

See Figure 3. Insert the four-foot piece of plastic pipe through both sides of the compression **T** (see the parts list in Table 1) and center it. Make 2-inch cuts in each end of the plastic pipe to allow it to clamp down on the poles. Put a 1-inch hose clamp loosely on each end of the pipe. Insert the butt end of each fishing pole approximately 6 inches into the plastic pipe and tighten the clamp. Assemble the remaining sections of the fishing pole. This will provide an assembly with 15.5 feet on each side of center, or 31 feet total. Put a sheet metal screw into the plastic pipe about 1 inch on each side of the **T** support. The screws will be used to attach the wire elements to the feed line.

Each fishing pole has a loop at the tip to guide the fishing line. I used it as a tie point for the end of each side of the dipole. Feed the #20 wire through the loop and twist a couple of turns to secure it. Wind the wire in a slow spiral for the full length of the element. This spiral forms a distributed loading coil so don't overdo it. I used about one turn per foot over the length of the cane pole and

Figure 1—The completed "Catfish Dipole" was tied to the chimney with nylon rope and was rotatable from ground level.

Figure 2—The center section of the dipole showing assembly details of the plastic pipe, compression T, mast, clamps and coaxial RF choke.

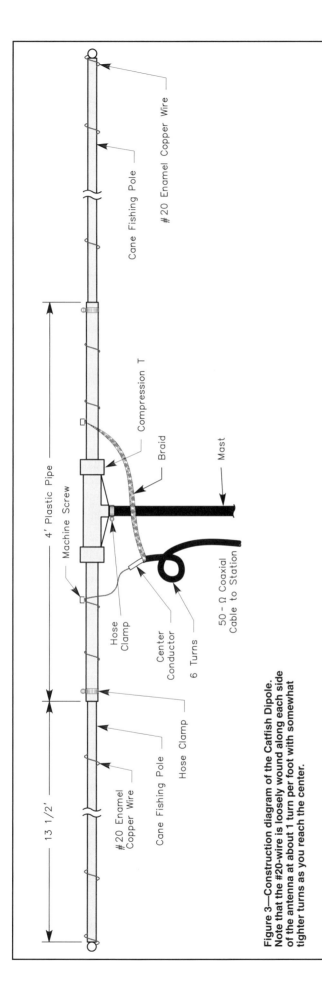

Figure 3—Construction diagram of the Catfish Dipole. Note that the #20-wire is loosely wound along each side of the antenna at about 1 turn per foot with somewhat tighter turns as you reach the center.

Table 1
Parts

Qty	Description	Cost
2	14-foot cane fishing poles	$ 9.90
2	25 feet of #20 enamel copper wire	$ 4.50
1	4 feet of 1-inch-OD schedule-40 plastic pipe	$ 2.00
1	Compression Ts (1 × 1 × $^3/_4$-inch pipe thread)	$ 2.95
1	Galvanized pipe nipple ($^3/_4$ × 3-inch)	$ 0.69
3	1-inch screw-type hose clamp	$ 2.25
1	(optional mast) 16-foot pool-cleaning pole (telescoping)	$19.95
Total cost		**$42.24**

added five closer-spaced turns around the plastic pipe on each side of center. Scrape the enamel from the wire and attach it under the sheet metal screw near the center **T**. Make the coax connections under the same two sheet metal screws and tighten. Do not overtighten. I removed the close-spaced turns (ultimately all of them) one at a time until the resonant frequency was 14.2 MHz.

Wind six or seven turns of the feed line coax into an RF choke near the feed point. This will decouple the unbalanced feed line from the balanced dipole. Figure 2 shows a poorly wound coaxial RF choke. I later rewound it on a four-inch cardboard form and taped it up for improved appearance and performance. Tape the coil securely to the mast and weatherproof the coax connections.

Performance

The expected input impedance at the center of a $^1/_2$-wavelength wire dipole is 72 Ω and should result in an SWR of 1.5:1 when fed with 50-Ω coax. I used the MFJ-259B Antenna Analyzer to measure the resonant frequency and SWR. The measured SWR at 14.2 MHz was 1.4:1, and didn't exceed 1.5:1 anywhere in the 20-meter band.

In the days after completion I made contacts with hams in Aruba, Slovenia and Chile—and even acted as net control station for the "Microcomputer Network" with coast-to-coast US stations, all with S-7 or better reports, while using a 100-W transceiver.

Future Plans

This antenna certainly served its purpose as an inexpensive temporary antenna, and it would be ideal for Field Day-type activities. For future projects, consider the following:

• Make the antenna even more portable by soldering the element wires to the metal ferrules that connect/separate the sections of the fishing pole. Taking the pole apart would also break down the wire elements.

• Wind multiband dipoles on the same pole (any frequency between 20 and 6 meters) and feed them from the same feed line.

• Build a half-size 40-meter dipole by adding loading coils on each side of the center plastic pipe.

• Add a suitably lightweight boom and another fishing pole assembly to make an inexpensive two-element Yagi.

• "Ruggedize" the design by switching to fiberglass fishing pole elements. You're moving into uncharted territory, but the thought is interesting, considering that inexpensive fiberglass poles are available at every Wal-Mart.

1513 Commanche Dr
Las Vegas, NV 89109
kd7va@arrl.net

By David Reid, PA3HBB

A Three Element Lightweight Monobander for 14 MHz

Not only is this portable antenna easy to build, it's light as a feather!

In preparation for the 2000 CQWW-CW contest for the PB6X Contest Group (**www.qsl.net/pb6x**), I started looking at my homemade 2-element 20-meter beam (see my Web site at **www.qsl.net/pa3hbb** for the article on this antenna). I decided that I needed more gain on 20 meters, along with a bit more front-to-back (F/B) ratio. But the beam had to be light and it should have the following qualities:
- easy to handle with one or possibly two people
- lightweight—but sturdy enough to handle the winter weather (always bad during a contest) and be built/taken down many times during a year
- reliable construction
- full size—to meet the F/B ratio and the forward gain required
- the ability to dismantle it easily for storage. I am not in a position to keep my antennas permanently erected because I live in a rented property.
- the ability to take the antenna into the field and on vacation.

Finding the Right Materials

With these goals in mind, I started looking into possible designs and materials to make the beam. Having designed and built a lot of beams in the past, I knew from experience that 3-element all-metal construction was possible. But to keep the elements from drooping too much and, mainly, to keep the weight down (and thus, the diameter/thickness/weight of the main boom), I ruled this option out at an early stage. I did explore the possibility of using metal elements, and performed some experiments; all of these proved that I was not going to meet all of my design criteria.

I had recently been experimenting with fiberglass fishing poles for making verticals, single-element delta loops and dipoles. So, I had a few left lying around the shack. Each of these was 6 meters long and extremely lightweight. "Perfect!" I said. "I have my elements. Now I just have to work out a way to mount them on a boom."

Again, experience held the solution. I opted for a piece of angle material made from aluminum, which is bolted to the main boom with two zinc-plated bolts at right angles to the boom. The zinc-plated bolts are important because if you use stainless steel, it will corrode the aluminum if you live in an environment where the air often carries a substantial salt content (near the ocean, for example).

I had done experiments with gain, SWR and front-to-back ratio on the 2-meter band a few years ago, so I dug out my notes and then scaled the dimensions to 20 meters.

But because I was planning to use wire for the elements (instead of $1/4$-inch tubing), I knew the diameter-to-wavelength ratio of the elements was going to be higher than the 2-meter equivalent. This meant that my wire elements had to be longer than the scaled design. The question was, how much longer?

To solve this problem, I first constructed an exact model of just the driven element from the same material I had used in my original research on the 2-meter model. I then scaled this to 20 meters, but replaced the tubing with the #14 copper wire. I knew it would be too short — but I also knew that if I measured the resonant frequency of the 20-meter wire version I could calculate how much longer I needed to make the final driven element.

As the whole antenna design is scaled, I could calculate the percentage of the difference and apply this percentage to the other elements. The spacing between the elements was going to change so minimally that I decided not to alter these dimensions.

Now I had the dimensions for the three elements: reflector, driven element and director. The spacing was a direct scaling from the 2-meter model.

I calculated the weight and wind loading for the antenna and, to see if my calculations were in the ballpark, I compared them to some commercial monoband antennas. My results were very favorable. I am by no means a mathematician, so I always make sure that my calculations are in the same region as other antennas. Now to build the prototype…

Designing the Prototype

With the lightweight fishing rods as the elements, I decided the boom could be much lighter than a beam with all metal elements. The boom was calculated to be 16 feet, 3 inches long. I made it from three 6-foot, 6-inch lengths of 1-inch × 2-inch extruded aluminum channel stock. The three boom sections were overlapped by 20 inches and two zinc-plated bolts were used in each section to bolt (2-inch) sides together in an overlapping fashion. See Figure 1.

This made a strong boom that could be dismantled into its original three pieces whenever necessary. The correct po-

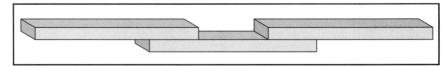

Figure 1—The boom sections.

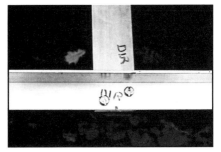

An angle section bolted to the boom.

Bill of Materials

6—20-foot fishing rods. If you have difficulty locating suitable fishing rods, substitute six SD-20 antenna supports from WorldRadio, 2120 28th St, Sacramento, CA 95818; tel 916-457-3655. $19 each plus $5 shipping and handling.
3—aluminum rectangular box sections, 1 × 2 inches for the boom.
3—1.2 × 1.2-inch sections of angle material for the element brackets.
6—2-inch bolts for attaching the angle material to the boom.
4—3-inch bolts to hold the boom sections together.
4—4-inch bolts to attach the boom to the mast plates.
1—14 × $1/4$-inch square printed circuit board for the boom-to-stub mast mounting plate.

sition of the elements was measured and marked on the boom and the three 3-foot, 3-inch pieces of angle aluminum were bolted to the boom sections at the appropriate places. These element bracket angles are held in place with two zinc-plated bolts each. See Figure 2.

The fishing rods were strapped to the angle material using three removable/adjustable zip-wraps per fishing rod. Once the elements were strapped to the angles, it was possible to determine the center of gravity of the beam in the middle of the garden and mark this on the boom (more about this later).

A short piece of 2×2 lumber was used as a temporary stub mast mounting. This was bolted to the boom using four metal plates with bolts going all the way through the boom and stub. (This was eventually replaced by two triangles of thick printed circuit board material.)

I raised my homemade mast and rested it on the fence surrounding my tennis court and then climbed a ladder with the antenna in one hand —it really *is* light and easy to handle—because the elements can stay telescoped while I am attaching the beam to the rotator.

Having put the boom (with the telescoped elements) onto the rotator, I extended all of the fishing rods and friction-locked them in place. I extended the reflector first, then rotated the antenna through 180° and extended the director. Finally, I extended the driven-element rods.

The last step was to raise the mast to the vertical position. It all seemed too easy. No problems were encountered and there was no time when I felt unsafe or unsteady on the ladder.

These experiments proved that it was possible to build the prototype mechanically, and it even looked like a real antenna. I left the antenna up for a week to see if it would suffer in the weather. We had some high winds and a lot of rain, but the antenna still stayed up and I was pleased when I took it down and found that all the parts were in perfect condition. It looked like I had a mechanical structure that would stand

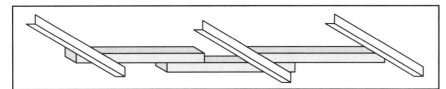

Figure 2—The boom and element brackets.

up for a lot more than just one weekend of heavy contesting.

The next step was to cut the wire elements, attach them to the fishing rods and put the whole antenna back up in the sky to see how it performs. Using the lengths I calculated earlier, I cut the #14 solid copper wire, marked the center point with tape and threaded it through the hoops on the fishing rods. I then taped the ends of the wire to the fishing rod so that tape at the center was sitting between the two rods at the centerline of the boom. Additionally, I secured the wire ends to each fishing rod with an extra zip-wrap fastener just to be sure they would stay in place.

In the prototype, the driven element was connected to the 50-Ω coaxial feed line through a 1:1 homemade balun, which allowed me to test the resonance of the beam

The antenna elements secured to the angle sections.

and determine the type of matching unit I required for the final antenna. A balun is generally necessary if you want your beam to have a directional pattern that is not distorted by the unbalanced feed line. However, it will also work without a balun. My preference is to use a balun on balanced antennas, but not on simple dipoles (or low beams such as my 2-element 80-meter wire beam, which is only 10 meters (33 feet) above the ground).

I assembled the beam again and put it back up on the mast. I connected my MFJ-259B antenna analyzer to the coaxial cable and the resonance was measured at 14.030 MHz and the impedance was 34 Ω. This was satisfactory. I could just use a 1:1 balun and still have an SWR of only 1.47:1. The 250-Hz 2:1 SWR bandwidth was about what I expected and it would certainly be sufficient for my needs as a CW-only antenna.

My first balun would not handle 400 W output, so a new one had to be built and tested. A 1:1.33 unun followed by a 1:1 balun would provide better match and Jerry Sevick, W2FMI, has some interesting designs in his book *Building and Using Baluns and Ununs*. But, because I am looking for a simple, lightweight design, I opted for the higher SWR and a simple 1:1 balun; my amplifier will easily load into 1.47:1.

With my first balun still on the antenna, I decided to check out the properties of the beam by listening on 20 meters to stations in different parts of the world using my Elecraft K2 QRP rig and rotating the beam to record the pattern, directivity and front-to-back ratio. Well, it acted like a beam; the front-to-back ratio was consistently over 20 dB. I compared the results against my 2-element 20-meter antenna, which has a front-to-back of approximately 12 dB and the

The author holds the finished antenna.

Temporary plates for the boom-to-sub mast.

3 element was always superior.

While the K2 was connected to the antenna, I could not resist calling CQ with the beam pointing Stateside. After a couple of calls I raised a few stations on the East Coast (while only running 3 W into the beam) and was getting 559 to 579 reports.

As far as forward gain goes, the antenna seemed to be quite a bit better than my 2-element antenna. Certainly I received better reports on the 3 element in every case.

Building the Antenna Yourself

If you'd like to duplicate my design, you'll be pleased to know that it is a simple matter of drilling the holes in the correct places and bolting the boom sections together, the angle sections to the boom and the mast mounting plates in place. The last step is to clamp the fishing poles onto the angle sections and secure the antenna element wires to the poles. If you have never built a Yagi antenna before, you should know that the driven element is essentially a dipole, so the wire must be cut into two equal halves and attached at the center to the feed line (in this case, to the two wires from the balun). See Figure 3.

The only tools required are a drill (with the right size of drill bits for the bolts), and an adjustable wrench to tighten the bolts. No cutting or bending or folding is required, making building the antenna easy even for less experienced amateurs. It also has another advantage when on vacation or in the field—only one tool is required for assembly (an adjustable wrench). The element dimensions are shown in Table 1. See the "Bill of Materials" sidebar for a list of the necessary parts. A drawing of the boom and element dimensions is available in Figure 4.

The fishing-pole supports for this antenna are a dielectric, so they actually lower the resonant frequency of the elements taped to them. There may be some variation in the exact dielectric properties of different brands of poles, so the antenna elements may need to be changed a bit. The director and reflector should not be very critical, so you can cut those to the lengths shown in Table 1. The driven element will be a bit more critical, so it may be necessary to add about 6 inches to the lengths shown and prune the length of the driven element until the antenna is resonant in your favorite part of the band. As designed, the SWR may be 2:1 at the point of best resonance.

The Balun

The 1-kW balun is made from a $2^1/_2$-inch diameter ferrite toroid with a permeability of 40, wound with 10 bifilar turns of #12 copper wire (Figure 3). The wires are taped together first, then wound onto the core. The windings are crossed through the core at the 50% point (5 turns) to allow easy connection of the coax to one end and the driven element wires to the other. The whole balun is mounted in a suitable plastic box to keep it out of the weather.

The Spacing Between the Elements

The spacing for the elements is a direct scaling from my 2-meter model and it provides a reasonable front-to-back gain and forward gain as well as an acceptable SWR (2:1 or less) for the transmitter.

The angle section for the reflector is bolted to one end of the boom at 90° to the boom. The driven element is placed at the end of this section of boom, 6 feet, 6 inches from the reflector (on the second section of the boom). The director is placed at the far end of the boom on the third section.

Finding the Center of Gravity

The next step was to find the center of gravity of the completed antenna. The boom and angle mounting brackets were ready for the elements (fishing poles) to be temporarily strapped in place. The antenna was assembled in the garden and I just picked up the beam and, using one hand, just kept moving my hand back and forth until the beam was stable and horizontal. When I found this point, I marked it as the beam center of gravity—the point where I wanted to fit the boom-to-mast clamps.

The Boom-to-Mast Clamp

There are several approaches you can use to secure the boom to the mast. One is shown in Figure 5. After several experiments with various materials, I wound up using plates made from printed circuit board material cut into triangles and bolted securely to the mast stub and the boom.

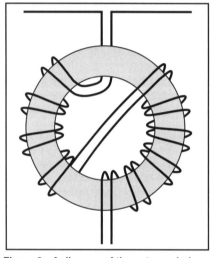

Figure 3—A diagram of the antenna balun.

Table 1
Element Dimensions

Note: All wire elements are composed of #14 solid copper wire.

Director	Driven Element	Reflector
31′, 6″	32′, 4″	35′, 10″

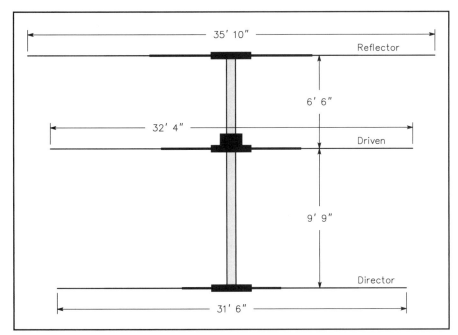

Figure 4—A drawing of the boom and element dimensions.

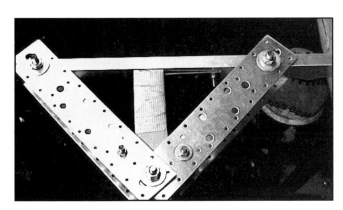

Figure 5—One approach to securing the boom to the mast.

Here you can see the wires on the elements themselves.

What Does it Weigh?

Traditionally, I weigh my antennas by putting the bathroom scales in the garden and, while holding the antenna, standing on the scales and recording the weight. Then I stand on the scales without the antenna and see the difference. With some quick subtraction I can determine the actual weight of the antenna. However, this method didn't work for this design—it was too light to measure the difference! So, I had to build a quick balance using a sawhorse and a long board, putting the beam on one end and weights on the other until it was stable and horizontal. According to my jury-rigged scale, the antenna only weighs 10 pounds!

Our thanks to Ed Hare, W1RFI, ARRL Laboratory Supervisor for his assistance in the preparation of this article. You can contact the author at Leenderweg 46, 5591 JE Heeze, The Netherlands; **pa3hbb@qsl.net**; **www.qsl.net/pa3hbb**.

By L. B. Cebik, W4RNL

A Simple Fixed Antenna for VHF/UHF Satellite Work

Explore the low-Earth orbiting amateur satellites with this effective antenna system.

When we are just getting interested in amateur satellite operation, the thought of investing in a complex azimuth-elevation rotator system to track satellites across the sky can stop us in our tracks. For starters, we need a simple, reliable, fixed antenna—or set of antennas—to see if we really want to pursue this aspect of Amateur Radio to its limit. We'll look at the basics of fixed antenna satellite work and develop a simple antenna system suited for the home workshop. There will be versions for both 145 and 435 MHz.

Turnstiles and Satellites

For more than decades, many fixed-position satellite antennas for VHF and UHF have used a version of the turnstile. The word "turnstile" actually refers to two different ideas. One is a particular antenna: two crossed dipoles fed 90° out of phase. The other is the principle of obtaining omnidirectional patterns by phasing almost any crossed antennas 90° out of phase. The first idea limits us to a single antenna. The second idea opens the door to adapting many possible antennas to omnidirectional work.

Figure 1 shows one general method of obtaining the 90° phase shift that we need for omnidirectional patterns. Note that the coax center conductor connects to only one of the two crossed elements. A $1/4$-λ section of transmission line that has the same characteristic impedance as the natural feed point impedance of the first antenna element alone connects one element to the next. The opposing ends of the two elements go to the braid at each end of the transmission line. If the elements happen to be dipoles, then a 70 to 75-Ω transmission line is ideal for the phasing line. However, the resulting impedance at the overall antenna feed point will be exactly half the impedance of one element alone. So we will obtain an impedance of about 35 Ω. For the dipole-based turnstile antenna, we'll either have to accept an SWR of about 1.4:1 or we'll have to use a matching section to bring the antenna to 50 Ω. A parallel set of RG-63 $1/4$-λ lines will yield about 43 Ω impedance, about right to bring the 35-Ω antenna impedance to 50 Ω for the main coax feed line. For all such systems, we must remember to account for the velocity factor of the transmission line, which will yield a line length that is shorter than a true quarter wavelength.

The dipole-based turnstile is popular for fixed-position satellite work. Figure 2 shows—on the left—one recommended system that has been in *The ARRL Antenna Book* since the 1970s. For 2 meters, a standard dipole-turnstile sits over a large screen that simulates ground. Spacing the elements from the screen by between $1/4$ and $3/8$ of a wavelength is recommended for the best pattern. For satellite operation, the object is to obtain as close to a dome-like pattern overhead as possible. The most desirable condition is to have the dome extend as far down toward the horizon as possible to let us communicate with satellites as long as possible during a pass.

The turnstile-and-screen system, while simple, is fairly bulky and prone to wind damage. However, the turnstile loses performance if we omit the screen. One way to reduce the bulk of our antenna is to find an antenna with its own reflector. However, it must have a good pattern for the desired goal of a transmitting and receiving dome in the sky. The dual Moxon rectangle array, shown in outline form on the right of Figure 2, offers some advantages over the traditional turnstile. First, it yields a somewhat better dome-like pattern. Second, it is relatively easy to build and compact to install.

Almost every fixed satellite antenna shows deep nulls at lower angles, and the number of nulls increases as we raise the antenna too high, thus defeating the desire for communications when satellites are at

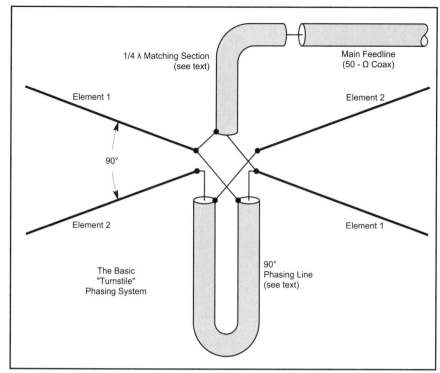

Figure 1—The basic turnstile phasing (and matching) system for any antenna set requiring a 90° phase shift between driven elements in proximity.

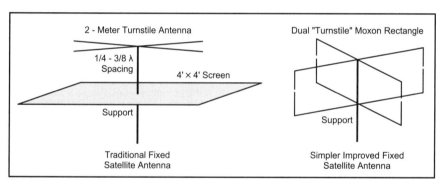

Figure 2—Alternative schemes for fixed-position satellite antennas: the traditional turnstile-and-screen and a pair of "turnstiled" Moxon rectangles.

be circular within under a 0.2-dB difference for 145.5 to 146.5 MHz, and within 0.5 dB for the entire 2-meter band. Since satellite work is concentrated in the 145.8 to 146.0 MHz region, the broadbanded antenna will prove fairly easy to build with success. A 435.6 MHz version, designed to cover the 435 to 436.2 MHz region of satellite activity will have an even larger bandwidth.

Like the dipole-based turnstile, the Moxons will be fed 90° out of phase with a $^1/_4$-λ phasing line of 50-Ω coaxial cable. The drivers will be connected just as shown in Figure 1. Since the natural feed-point impedance of a single Moxon rectangle of the design used here is 50 Ω, the pair will show a 25-Ω feed-point impedance. Paralleled $^1/_4$-λ sections of 70- to 75-Ω coaxial cable will transform the low impedance to a good match for the main 50-Ω coaxial line to the rig. In short, we have "turnstiled" the Moxon rectangles into a reasonable fixed-position satellite antenna.

Building the Moxon Pairs

The Moxon rectangle is a modification of the reflector-driver Yagi parasitic beam. However, instead of using linear elements, the driver and reflector are bent back toward each other. The coupling between the ends of the elements combined with the coupling between parallel sections of the elements combine to produce a pattern with a broad beamwidth. By carefully selecting the dimensions, we can obtain both good performance (meaning adequate gain and an excellent front-to-back ratio) and a 50-Ω feed point impedance.[1]

In fact, a single Moxon rectangle might be used on each band for reasonably adequate satellite service. When pointed straight up, the Moxon rectangle pattern is a very broad oval, although not a circle. The oval pattern also gives the Moxon another advantage over dipoles in a turnstile configuration. If the phasing-line between dipoles is not accurately cut, the normal turnstile near-circle pattern degrades into an oval fairly quickly because the initial single dipole pattern is a figure **8**. The single Moxon oval pattern allows both dimensional inaccuracies and phasing-line inaccuracies of considerable amounts before degrading from a nearly perfect circle.

Figure 5 shows the critical dimensions for a Moxon rectangle. The lettered references are keys to the dimensions

low angles. Figure 3 shows the elevation patterns of a turnstile-and-screen and of a pair of Moxon rectangles when both are 2λ above the ground. A 1λ height will reduce the low angle ripples even more, if that height is feasible. However, the builder always has to balance the effects of height on the pattern against the effects of ground clutter that may block the horizon.

The elevation patterns show the considerably smoother pattern dome of the Moxon pair over the traditional turnstile. The middle of the turnstile dome has nearly 2 dB less gain than its peaks, while the top valleys are nearly 3 dB lower than the peaks. The peaks and valleys can make the difference between successful communications and broken-up transmissions. So, for the purpose of obtaining a good dome, the Moxon pair may be superior.

A reasonable suggestion offered to me was simply to add reflectors to a standard dipole turnstile and possibly obtain the same freedom from a grid or screen structure. Figure 4 shows the limitation of that solution. The result of placing reflectors behind the dipole turnstile is a pair of crossed 2-element Yagi beams fed 90° out of phase. The pattern is indeed circular and stronger than that of the Moxon pair. However, the beamwidth is reduced to only 56° at the half-power points. The antenna would make an excellent starter for a tracking AZ-EL rotator system, but it does not have the beamwidth for good fixed-position service.

The Moxon pair, with lower but smoother gain across the sky dome, offers the fixed-antenna user the chance to build a successful beginning satellite antenna. The pattern will

---

[1]See "Having a Field Day with the Moxon Rectangle," *QST*, June, 2000, pp 38-42, for further details on the operation of the Moxon rectangle, along with the references in the notes to that article. Also included in the notes is the source for a program to calculate the dimensions for a 50-Ω Moxon rectangle for any HF or VHF frequency using only the design frequency and the element diameter as inputs.

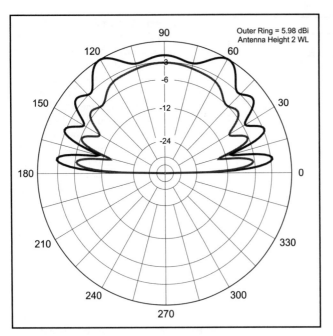

Figure 3—A comparison of elevation patterns for the turnstile-and-screen system (with $3/8\lambda$ wavelength spacing, shown in blue) and a Moxon pair (shown in red), both at 2λ height.

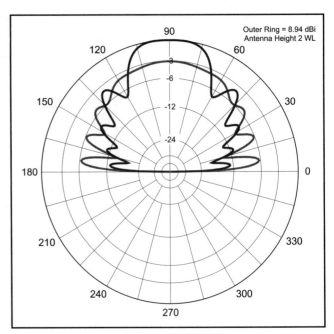

Figure 4—A comparison of elevation patterns for 2-element turnstiles (crossed 2-element Yagis, shown in blue) and a Moxon pair (shown in red), both at 2λ height.

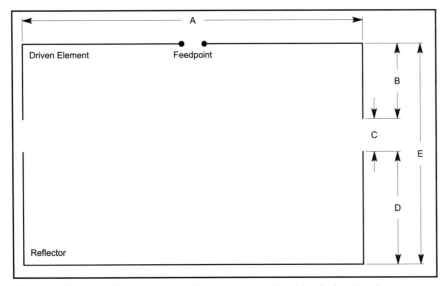

Figure 5—The basic dimensions of a Moxon rectangle. Two identical rectangles are required for each "turnstiled" pair.

Table 1

Dimensions for Moxon Rectangles for Satellite Use

Two are required for each antenna. The phase-line is 50-Ω coaxial cable and the matching line is parallel sections of 75-Ω coaxial cable. Low power cables less than 0.15 inches in outer diameter were used in the prototypes. See Figure 5 for letter references. All dimensions are in inches.

Dimension	145.9 MHz	435.6 MHz
A	29.05	9.72
B	3.81	1.25
C	1.40	0.49
D	5.59	1.88
E (B + C + D)	10.80	3.62
$1/4$ wavelength	20.22	6.77
0.66 velocity factor phasing and matching lines	13.35	4.47

in Table 1. The design frequencies for the two satellite antenna pairs are 145.9 MHz and 435.5 MHz, the centers of the satellite activity on these two bands. The 2-meter Moxon prototype uses $3/16$-inch diameter rod, while the 435 MHz version uses #12 AWG wire with a nominal 0.0808-inch diameter. (Single Moxons built to these dimensions would cover all of 2-meters and about 12 MHz of the 432 MHz band.) Going one small step up or down in element diameter will still produce a usable antenna, but major diameter changes will require that the dimensions be recalculated.

The reflectors are constructed from a single piece of wire or rod. I use a small tubing bender to create the corners. The rounding of the corners creates a slight excess of wire for the overall dimensions in the table. I normally arrange the curve so that the excess is split between the side-to-side dimension (A) and the reflector tail (D). Practicing on some scrap house wire may make the task go well the first time with the actual aluminum rod. The total reflector length should be A + (2 × D).

The driver consists of two pieces, since we'll split the element at its center for the feeding and phasing system. I usually make the pieces a bit longer before bending and trim them to size afterwards. The total length of the driver, including the open area for connections, should be A + (2 × B).

Perhaps the most critical dimension is the gap, C. I have found nylon tubing, available at hardware depots, to be very good to keep the rod ends aligned and correctly spaced. When everything has been tested and found correct, a little super-glue on the tubing ends and aluminum stands up to a lot of wind. I usually nick the aluminum just a little to let the glue settle in and lock the junction. For the UHF version, a short length of heat-shrink tubing provides a lock

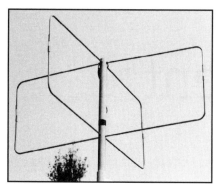

A close-up view of the 145.9 MHz rectangle pair.

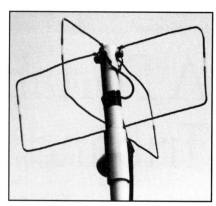

The 435-MHz Moxons.

The feed point assemblies are attached to solder lugs. The phasing line is routed down one side of the support, while the matching section line is run down the other. Electrical tape holds them in place. For worse weather, the tape may be over-sealed with butylate or other coatings. Likewise, the exposed ends of the coax sections and the contacts themselves should be sealed from the weather. The details can be seen—as built for the experimental prototypes in one of the photos—before sealing, since lumps of butylate or other coatings tend to obscure interesting details.

The overall assembly of the two antennas appears in the second photograph. The PVC from the support **T**s can go to a center Tee that also holds the main support for the two antennas. A series of adapters, made from miscellaneous PVC parts to fit over a standard length of TV mast. Alternatively, the antennas can be separately mounted about 10 feet apart. The 10-foot height of the assembly has proven adequate for general satellite reception, although I live almost at the peak of a hill.

The antennas can be mounted on the same mast. However, for similar sky-dome patterns, they should each be the same number of wavelengths above ground. For example, if the 2-meter antenna is about two wavelengths up at about 14 feet or so, then the bottom of the 435-MHz antenna should be only about 4.5 feet above the ground. Placing the higher-frequency antenna below the 2-meter assembly will create some small irregularities in the desired dome pattern, but not serious enough to affect general operation.

There is no useful adjustment to these antennas except for making the gap between the drivers and reflectors as accurate as possible. Turnstile antennas show a very broad SWR curve. Across 2 meters, for example, the highest SWR is under 1.1:1. However, serious errors in the phasing line length can result in distortions to the desired circular pattern. There is no substitute for checking the lengths of the phasing line and the matching section several times before cutting. The correct length is from one junction to the next, including the portions of exposed cable interior.

These two little antennas will not compete with tracking AZ-EL rotating systems for horizon-to-horizon satellite activity. For satellite work, however, power is not always the problem (except for using too much) and modern receiver front-ends have enough sensitivity to make communication easy. So when the satellite reaches an angle of about $30°$ above the horizon, these antennas will give a very reasonable account of themselves. When you become so addicted to satellite communication that you invest in the complete tracking system, these antennas can be used as back-ups while parts of the complex system are down for maintenance!

You can contact the author at 1434 High Mesa Dr, Knoxville, TN 37938; **cebik@cebik.com**.

for the size of the gap and the alignment of the element tails.

It is one thing to make a single Moxon and another to make a working crossed pair. Figure 6 shows the general scheme that I used for the prototypes, using CPVC. (Standard schedule 40 or thinner PVC or fiberglass tubing can also be used.) The support stock is $3/4$ inch nominal. The reflectors go into slots at the bottom of the tube and are locked in two ways. Whether or not the two reflectors make contact at their center points makes no difference to performance, so I ran a very small sheet screw through both 2-meter reflectors to keep their relative positions firm. I soldered the centers of the 435-MHz reflectors. Then I added a coupling to the bottom of the CPVC to support the double reflector assembly and to connect the boom to a support mast. Cementing or pressure fitting the cap is a user option.

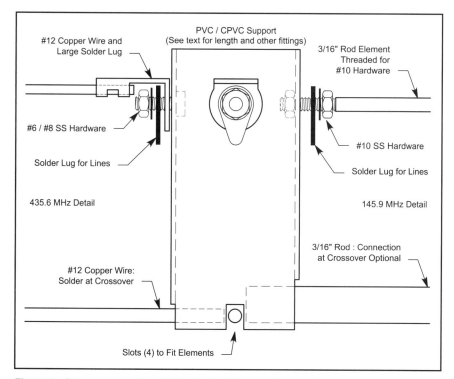

Figure 6—Some construction details for the Moxon pairs constructed as prototypes.

By Markus Hansen, VE7CA

A Portable 2-Element Triband Yagi

Have you ever dreamed about a portable beam you could use at your summer cottage, while camping or on Field Day? Dream no longer. This portable beam can be rolled up and stashed in your car's ski boot!

Several years ago I entered the ARRL November Sweepstakes CW contest in the QRP category, operating from a portable location. It turned out to be a very frustrating experience with only 3 W of output power and dipole antennas. After the contest I decided that the next time I entered a QRP contest it had to be with gain antennas.

My philosophy has always been to try to keep life as simple as possible. In other words, I look for the easiest way to accomplish a goal that guarantees success. Don't get me wrong: Dipoles work particularly well considering the time and effort put into making them. But adding a reflector to a dipole antenna increases the overall gain about 5 dB, depending on the spacing between the elements. This extra gain makes a significant difference, especially when you are dealing with QRP power levels. My 3-W transmitted signal would sound like a 9.5-W powerhouse just by adding another piece of wire! And it would be inexpensive too.

With Solar Cycle 23 in full swing, having an antenna with gain on 15 and 10 meters also became a consideration. Another parameter was the sale of the family van, which meant the new antenna had to fit into the ski boot of our car. Keeping these constraints in mind, I used a computer antenna-modeling program, trying different design parameters to develop a triband 2-element

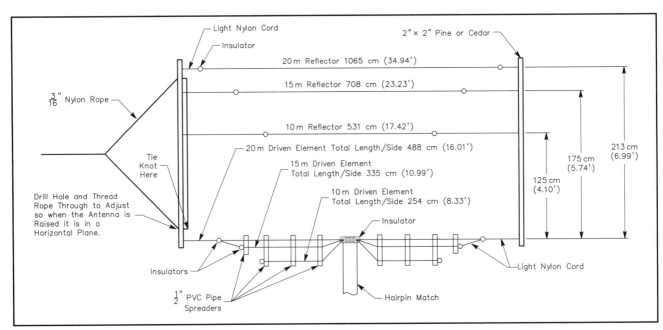

Figure 1—Dimensions for VE7CA's 2-element wire triband Yagi.

portable Yagi using wire elements.

The basic concept comprises three individual dipole driven elements, one each for 10, 15 and for 20 meters tied to a common feed point, plus three separate reflector elements. The elements are strung between two 2.13-meter (7-foot) long, 2×2-inch wood spreaders, each just long enough to fit into the ski boot of the car. Use the lightest wood possible, such as cedar, pine or spruce to keep the total weight of the antenna as light as possible. Fiberglass poles would also work, or PVC pipe reinforced with maple doweling to ensure they don't bend. (Wood has the benefit of being easy-to-find and very affordable).

Adding a reflector element relatively close to the driven elements lowers the feed-point impedance of the driven element, so a simple hairpin match was employed to match the driven elements to a 50-Ω feed line. Figure 1 shows the layout and dimensions of the antenna.

The Hairpin Match

The matching system is very simple and foolproof. You should be able to copy the dimensions shown in Figure 2 and not need to retune the hairpin match, unless you plan to use the antenna in the top portions of the phone bands. The dimensions in Figure 2 produced a very low SWR—under 1.3:1 over the CW portions of all three bands. However, even in the lower portions of the SSB bands, the SWR doesn't rise above 2:1. SWR measurements were made at the end of a 25-meter (82-foot) length of RG-58 coax feed line.

Some may wonder why I used such a long feed line. First, when operating from a portable location it is better to be long than short. Nothing is more frustrating than finding that the coax you took along with you is too short. Further, when I change beam direction I walk the antenna around the antenna support, thus requiring a longer length than if I went directly from the antenna to the operating position.

If you are concerned about line loss you can run RG-58 down to the ground and larger-diameter RG-8 or RG-213 to the operating position. You may also find that in your particular situation a shorter length of coax will do. An 18-meter (59-foot) long piece of RG-58 has a loss of about 1 dB at 14 MHz, which is entirely acceptable considering the convenience of using coax cable.

Adjusting the Hairpin Match

If after raising the antenna the SWR is not as low as you want in the portion of the bands you plan to operate, first double-check to make sure that all the elements are cut to the correct length and that the spacings between the driven elements and reflectors are correct. Next you can adjust the hairpin match. Connect either an antenna SWR analyzer or a transmitter and SWR meter to the end of the feed line and pull the antenna up to operating height. Determine where the lowest SWR is on 15 meters. By moving the shorting bar on the hairpin match up or down you can adjust the lowest SWR point to the middle of the portion of the 15 meter band you prefer. If your preference is near the top end of 15 meters you may have to shorten the 15-meter driven element slightly. After adjusting the 15-meter element and hairpin match, adjust the 10 and 20 driven-elements lengths separately, without changing the position of the shorting bar on the hairpin match.

The hairpin match is very rugged. You can attach the feed line to it with tape, roll it up, pack the antenna away and even with the matching wires bent out of shape it just seems to want to work.

Antenna Support

Adhering to my constraint to keep things as simple as possible, I only use one support for the antenna, typically a tree. When the antenna is raised to its operating position it is a sloping triband Yagi. To achieve this, attach a rope to each end of the 2×2's to form a V-shaped sling, as shown in the Figure 1. Attach a length of rope to one sling and pull the antenna up a tree branch, tower or whatever vertical support is available. Tie a second length of rope to the bottom sling and anchor the antenna to a stake in the ground. By putting in two or three stakes in the ground around the antenna support, you can walk the antenna around to favor a particular direction. To change direction 180°, give the feed line a pull and the array will flip over. So simple but very effective!

Local or DX

One of the features of a sloping antenna is that you can adjust the take-off (elevation) angle. For example, if you are interested in North American contacts (whether for casual QSOs or the ARRL SS contest), then sloping the antenna away sideways from the support structure at 45° with the feed point approximately 8 meters (26 feet) above the ground, will yield a 20-meter pattern similar to Figure 3A. Here, the maximum lobe is between 10° and 60° in elevation. The pattern of the antenna in a flat-top horizontal configuration at 9.1 meters (30 feet) is overlaid for comparison. You can see that the tilted beam has better low-angle performance, but at higher angles has less gain than its horizontal counterpart. Figure 3B shows an overlay of the azimuth patterns for these two configurations at a 10° takeoff angle.

If DX is your main interest, then you want to position the antenna even closer to vertical to emphasize the lower elevation angles. Figure 4 shows the pattern on 20 meters when the antenna is tilted sideward 10° away from vertical, again compared with the other orientations in Figure 3A. The feed point is 6 meters above ground and the model assumes fresh water in the far field, which is the case at my portable location.

Remember that the radiation pattern is quite dependent on ground conductivity and dielectric constant for a vertically polarized

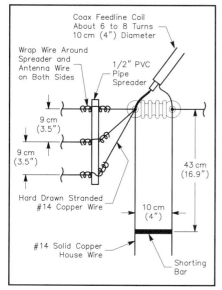

Figure 2—Close-up view of the feed point.

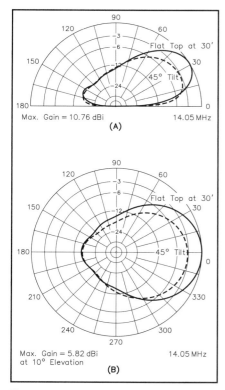

Figure 3—At A, comparison of elevation patterns for VE7CA Yagi as a horizontal flat top (solid line) and tilted 45° from vertical (dashed line). At B, comparisons of azimuth patterns for a 10° elevation angle.

53

antenna. A location close to saltwater will yield the highest gain and the lowest radiation angle. With very poor soil in the near and far field, the peak radiation angle will be higher and the gain less.

I have had the opportunity to test this out at my portable location. Using two trees as supports, I am able to pull the antenna close to horizontal with the feed point about 7 meters above the ground. In this position, with 20 meters open to Europe, I have found it difficult to work DX on CW with 3 W of output power. However, when I change the slope of the antenna so that it is nearly vertical I not only hear more DX stations, but I find it relatively easy to work DX.

I have tried this many times, since it is simple to lower one end of the antenna to change the slope and hence the radiation take-off angle. The sloping antenna always performs much better for working DX than a low horizontal antenna. Recently, I worked nine European countries during two evenings of casual operating, even though the highest end of the antenna was only about 10 meters high, limiting the slope to about 45°.

Figure 5 shows the elevation pattern on 28.05 MHz for the beam sloped 10° from vertical at 45° from vertical, with the feed point at 8 meters height, again compared with the beam as a flat top at 9.1 meters (30 feet). With a steeper vertical slope, the 10-meter elevation pattern has broken into two lobes, with the higher-angle lobe stronger than the desired low-angle lobe.

This demonstrates that it is possible to be too high above ground for a vertically polarized antenna. Lowering the antenna so that the bottom wires are about 2.5 meters (8 feet) above ground (for safety reasons) restores the 10-meter elevation pattern without unduly compromising the 20-meter pattern.

Portable It Is

A winning feature of this antenna is that it is so simple to put up, take down, transport and store away until it is needed again. When I am finished using the antenna and it's time to move on, I just lower the array and roll the wire elements onto the 2×2's. I put a plastic bag over each end of the rolled-up array and tie the bag with string so that the wires don't come off the ends of the 2×2's. I then put it in the ski boot of a car, or in the back of a family van and away we go. At home, it takes very little space to store and it is always ready to go—No bother, no fuss.

Testimonial

How well does it work? It works very well. On location I use a bow and arrow to shoot a line over a tall tree and then pull one end of the array up as far as possible. For DX I aim for a height of 20 to 30 meters if possible. For the Canada Day, Field Day and Sweepstakes contests I aim for a height of about 15 meters. This antenna helped me to achieve First Place for Canada, in the 1997 ARRL CW Sweepstakes Contest, QRP category.

The ability to quickly change direction 180° is a real bonus. Late in the 1997 ARRL SS CW contest with the antenna pointed east I tuned across KH6ND. He was the first Pacific station I had heard during the contest and obviously I needed to work him. After trying many times to break through the pileup and not succeeding, I flipped the antenna over to change the direction 180° and then worked him on my next call. Figure 6 shows the azimuth pattern at 21.05 MHz for the beam mounted with a 10° slope from vertical. There is a very slight skewing of the azimuthal pattern because the slope away from purely vertical makes the antenna geometry asymmetrical.

VE7NSR, the North Shore Amateur Radio Club, has used this antenna sloped at about 45° for the last two years on 20 and 10 meters on Field Day with good success. The title photo shows the antenna attached to a tower during Field Day.

As they say, the proof is in the pudding. If you need a 20 to 10 meter antenna with gain, this has to be one of the simplest antennas to build, and it will work every time!

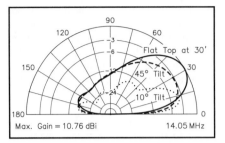

Figure 4—Comparison of elevation patterns for VE7CA Yagi as a horizontal flat top (solid line), tilted 45° from vertical (dashed line) and tilted 10° from vertical (dotted line).

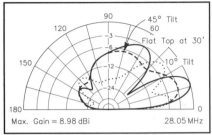

Figure 5—Same antenna configurations as shown in Figure 4, but at 28.05 MHz. On 10 meters, the flattop configuration is arguably best, but the 45° tilted configuration is not far behind.

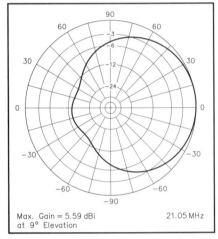

Figure 6—Azimuthal pattern for VE7CA Yagi tilted 10° from vertical on 15 meters.

Markus Hansen, VE7CA, was first licensed as VE7BGE in 1959. He has been a member of ARRL since he received his license. His main interests include DX, collecting grids on 6 meters, contesting and building his own antennas and various types of ham-radio equipment. He is also an ardent CW operator. Markus has had two previous articles published: "The Improved Telerana, with Bonus 30/40 Meter Coverage," in The ARRL Antenna Compendium Vol 4 *and "Two Portable 6-Meter Antennas" in* The ARRL Antenna Compendium Vol 5. *You can contact Markus at 674 St Ives Cres, North Vancouver, BC V7N 2X3, Canada, or by e-mail at* ve7ca@rac.ca.

You can download the EZNEC input-data files as **VE7CA-1.ZIP** *from ARRLWeb* (**www.arrl.org/files/qst-binaries/**).

By Rich Wadsworth, KF6QKI

A Portable Twin-Lead 20-Meter Dipole

With its relatively low loss and no need for a tuner, this resonant portable dipole for 14.060 MHz is perfect for portable QRP.

My first attempt at a portable dipole was using 20 AWG speaker wire, with the leads simply pulled apart for the length required for a $^1/_2$ wavelength top and the rest used for the feed line. The simplicity of no connections, no tuner and minimal bulk was compelling. And it worked (I made contacts)!

Jim Duffey's antenna presentation at the 1999 PacifiCon QRP Symposium made me rethink that. The loss in the feed line can be substantial, especially at the higher frequencies, if the choice in feed line is not made rationally. Since a dipole's standard height is a half wavelength, I calculated those losses for 33 feet of coaxial feed line at 14 MHz. RG-174 will lose about 1.5 dB in 33 feet, RG-58 about 0.5 dB, RG-8X about 0.4 dB. RG-8 is too bulky for portable use, but has about 0.25 dB loss. For comparison, *The ARRL Antenna Book* shows No. 18 AWG zip cord (similar to my speaker wire) to have about 3.8 dB loss per 100 feet at 14 MHz, or around 1.3 dB for that 33 feet length. Note that mini-coax or zip cord has about 1 dB more loss than RG-58. Are you willing to give up that much of your QRP power and your hearing ability? I decided to limit antenna losses in my system to a half dB, which means I draw the line at RG-58 or equivalent loss.

TV Twin Lead

It is generally accepted that 300 ohm ribbon line has much less loss than RG-58. Some authors have stated that TV twin lead has similar loss as RG-58, which is acceptable to me. A coil of twin-lead is less bulky and lighter than the same length of RG-58. These qualities led me to experiment with it. One problem is that its 300 ohm impedance normally requires a tuner or 4:1 balun at the rig end.

But, since I want approximately a half wavelength of feed line anyway, I decided to experiment with the concept of making it an exact electrical half wavelength long. Any feed line will reflect the impedance of its load at points along the feed line that are multiples of a half wavelength. Since a dipole pitched as a flat-top or inverted V has an impedance of 50 to 70 ohms, a feed line that is an electrical half wave long will also measure 50 to 70 ohms at the transceiver end, eliminating the need for a tuner or 4:1 balun.

To determine the electrical length of a wire, you must adjust for the velocity factor (VF), the ratio of the speed of the signal in the wire compared to the speed of light in free space. For twin lead, it is 0.82. This means the signal will travel at 0.82 times the speed of light, so it will only go 82% as far in one cycle as one would normally

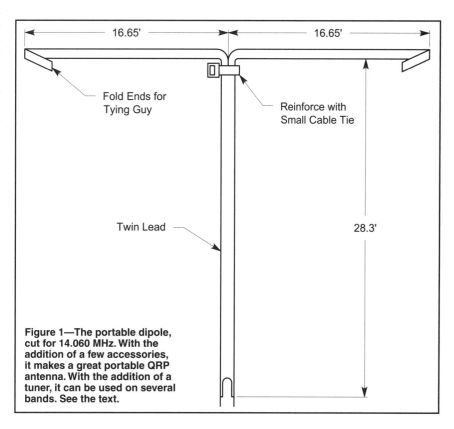

Figure 1—The portable dipole, cut for 14.060 MHz. With the addition of a few accessories, it makes a great portable QRP antenna. With the addition of a tuner, it can be used on several bands. See the text.

55

compute using the formula 984/f(MHz). I put a 50 ohm dummy load on one end of a 49 ft length of twin lead and used an MFJ 259B antenna analyzer to measure the resonant frequency, which was 8.10 MHz. The 2:1 SWR bandwidth measured 7.76 to 8.47 MHz, or about 4.4% from 8.10 MHz.

The theoretical $1/2$ wavelength would be 492/8.1 MHz, or 60.7 feet, so the VF is 49/60.7=0.81, close to the 0.82 that is published. A $1/2$ wave for 14.06 MHz would therefore be 492×0.81/14.06 or 28.3 feet. I cut a piece that length, soldered a 51 ohm resistor between the leads at one end, and hoisted that end up in the air. I then measured the SWR with the 259B set for 14.060 MHz and found it to be 1:1. I used the above-measured 2:1 bandwidth variation of 4.4% to calculate that the feed line could vary in length between 27.1 and 29.5 feet for a 2:1 maximum SWR.

Now comes the fun part. With another length of twin lead, I cut the web between the wires, creating 17 ft legs, and left 28.3 feet of feed line. I hung it 30 feet high, tested, and trimmed the legs until the 259B measured 1:1 SWR. The leg length ended up at 16.75 feet. (Note: The VF determined above only applies to the feed line portion of the antenna.) There is no soldering and no special connections at the antenna feed point. I left the ends of the legs an inch longer to have something to tie to for hanging. I reinforced the antenna end of the uncut twin lead with a nylon pull tie, with another pull tie looped through it to tie a string to it for using as an inverted V. To connect the feed line to the transceiver, I used a binding post-BNC adaptor that is available from Ocean State.[1] My original intention of leaving the feed line free of a permanent connector was to allow connection to an Emtech ZM-2 balanced antenna binding post connectors. Since then I have permanently attached a short stub of RG-58 with a BNC, because I plan to either use it with my single band 20 meter Wilderness Radio SST, or with an Elecraft K1 or K2 with built-in tuner. I did this by connecting the shield to one side of the twin lead and the center conductor to the other side—no balun was used between the coax and twin lead.

After a year or so of use and further field testing, including different heights and V angles, I further trimmed the legs to a length of 16.65 feet. I found that the lowest SWR was usually obtained with the V as close to 90 degrees as I could determine visually. Also, I found that the resonant frequency (or at least the frequency at which SWR

[1]Ocean State Electronics, 6 Industrial Dr, Westerly RI 02891, tel 800-866-6626 or 401-596-3080; fax 401-596-3590; e-mail: **ose@oselectronics.com**.

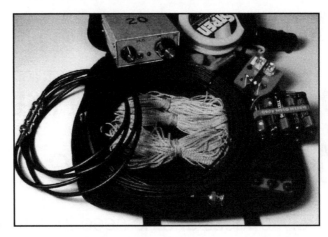

Figure 2—The author's portable station, including twin-lead dipole, 20-meter Wilderness Radio SST transceiver and support line. It all fits in the 8"×10$1/2$"×2" Compaq notebook computer case.

was at a minimum) is lower if the antenna is closer to the ground, and vice-versa. For example, with the top of the V at 22 feet, the lowest SWR was measured at around 13.9 MHz, and with the top of the V at 31 feet, SWR was lowest at around 14.1 MHz. In both cases, SWR at 14.060 did not exceed 1.3:1.

I used Radio Shack 22 AWG twin lead that is available in 50 ft rolls. To have no solder connections, you need at least 45 feet. When I cut the twin lead to make the legs, I just cut the "web" down the middle and didn't try to cut it out from between the wires. It helps make the whole thing roll up into a coil, and the legs don't tangle when it's unrolled, since they're a little stiff. It turned out that the entire antenna is lighter than a 25 ft roll of RG-58. This antenna can be scaled up or down for other frequencies also. An even lower loss version can be made with 20 AWG 300 ohm "window" line, though the VF of that line is different and should be measured before construction.

How High?

Wait, you say—"After all that talk about having it a half wave up, you only have it up 28 feet." A 6 or 12 ft RG-58 jumper, available with BNC connectors from RadioShack, can be used to get it higher if the right branch is available. Since impedance at the feed point is 50 to 70 ohms, 50 ohm coax can be used to extend the feed line. I have used it in the field a few times as an inverted V, at various heights and leg angles, and used an SWR meter to double-check its consistency in different situations. SWR never exceeded 1.5:1, so I feel safe leaving the tuner home. For backpacking, I leave the SWR meter home, too!

And there's a bonus: Since it has a balanced feed line, it *can* be used with little loss as a multi-band antenna, with a tuner, from 10 to 40 meters. I quote John Heyes, G3BDQ, from *Practical Wire Antennas*, page 18:

"Even when the top of the doublet antenna is a quarter-wavelength long, the antenna will still be an effective radiator." Heyes used an antenna with a 30 ft top length about 25 ft off the ground on 40 meters and received consistently good reports from all Europe and even the USA (from England). It will not perform as well at 40 meters as at 20 meters, however, though 10 through 20 meters should be excellent.

Testing, Testing

To test this theory, I recently worked some of Washington State's Salmon Run contesters and worked many Washington hams and an Ohio and a Texas station on 15 and 20 meters, with the antenna up 22 feet on a tripod-mounted SD20 fishing pole, using 10 W from an Elecraft K2 from central California. The K2 tuner was used to tune the antenna on 15 meters. Signal reports were from 549 to 599. Unfortunately, this was a daytime experiment and 40 meters was limited to local traffic.

At the 2001 Freeze Your Buns Off QRP contest, it was hung at 30 feet and compared to a 66 ft doublet up 50 feet on 10, 15 and 20 meters, using a K2 S-meter. There was little if any difference. At the 2001 Flight of the Bumblebees QRP contest I compared it, at 20 meters, to a resonant wire groundplane antenna with each antenna top at 20 feet and found it to consistently outperform the groundplane. I have concluded through these informal experiments that a resonant inverted V, when raised at a height close to or exceeding a half wavelength, produces the most "bang for your buck" and that extra length or height beyond that yields diminishing returns.

A ham since 1998, Rich Wadsworth, KF6QKI, is a civil engineer in private practice as a consultant. Since earning his license, he reports, that he has become obsessed with kits and homebrewing. You can reach Rich at 320 Eureka Canyon Rd, Watsonville, CA 95076; **richwads@compuserve.com**.

By Rick Rogers, KI8GX

A "One-Masted Sloop" for 40, 20, 15 and 10 Meters

What started off as a compromise replacement for a "monster loop" turned out much better than expected. This antenna may prove to be an exception to the rule that "you get what you pay for."

Over 33 years of hamming, one of my favorite activities is building and testing antennas. Of all the types of antennas tried, I get the best bang for the buck from simple, horizontal loops.

Designing the Loop

An interesting property of loop antennas is that they are harmonically resonant. As shown by Doug DeMaw, W1FB, a loop designed for 7.1 MHz will also resonate at 14.2 MHz, 21.3 MHz, 28.4 MHz, etc.[1] See Figure 1. The ability to operate on multiple bands without retuning and the multidirectional nature of their radiation patterns make horizontal loops especially useful for DX, contest, and net control applications where having to wait to rotate a beam can be a disadvantage. Another advantage of the loop antenna is that it tends to be quieter on receive than some other designs, such as Yagis or verticals.[2]

The best antenna I ever built was a 160-meter full-wave horizontal loop. Even though the antenna was only up about 35 feet, it did a pretty good job on 160, is spite of radiating most of its energy skyward. Where this antenna was really effective, though, was on its harmonics. An EZNEC[3] model of this antenna shows, for example, that at 10 meters, it radiates multiple low-angle lobes, some with gain figures of more than 13 dBi.

Of course, a monster like this had (note past tense!) its problems. It required 4 masts, 540 feet of wire and a big chunk of land. As the reader might guess, antennas that big suffer a lot from the wind, even if made out of relatively strong wire. Mine was made of 17 gauge aluminum fence wire but it seemed like I was always repairing damaged masts and broken wires. [Solid wire is more likely than stranded wire to break as a

[1] Notes appear on page 59.

result of repeated flexing.—Ed.] After about six months of constant struggle against the elements, the antenna and three of its four supports succumbed to wind-driven hail.

After the storm, and several unsatisfying weeks trying to get by with a home-brew vertical, I thought to try something a little less ambitious. What I had in mind was a loop that would use only the single remaining support. A quick session with EZNEC showed that a sloping loop, 140 feet in circumference (a full wavelength on 40 meters), with the feed point elevated on a single 30-foot support should resonate on 40, 20, 15 and 10 meters. The antenna should also produce reasonable gain in multiple directions, especially at the shorter wavelengths (see Figure 2). This "one-masted sloop," a *sloping loop* supported and fed at the top corner, turns out to be a good performer and costs almost nothing.

Building the Loop

Construction couldn't be easier. First, buy or build a dipole center insulator with coaxial connector as described in *The ARRL Handbook for Radio Amateurs* (see Figure 3).[4] Connect the opposite ends of the 140-foot wire to the center insulator. I prefer 14 gauge stranded and insulated wire because it is easy to work with. Tie 50-foot lengths of $3/16$-inch rope to the antenna at points

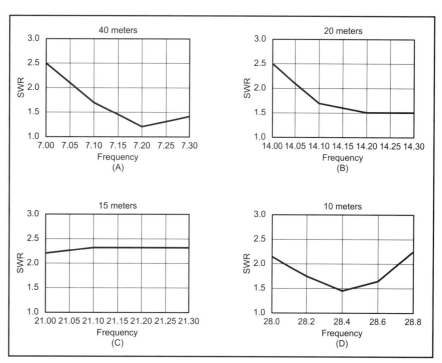

Figure 1—SWR vs. frequency plots for the 136-foot, 40 through 10-meter sloop. The SWR minimum for the four bands is easily adjusted by adding or deleting small lengths of wire from the loop.

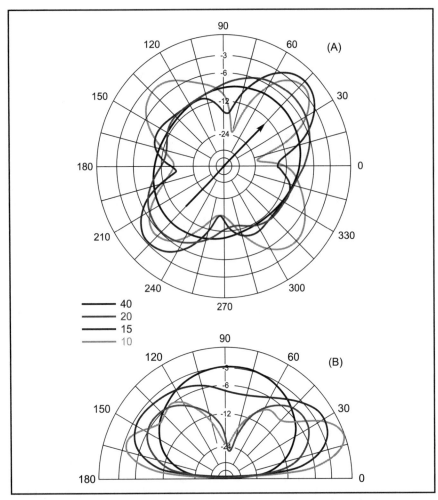

Figure 2—*EZNEC* study of the far field radiation patterns of the 40 through 10-meter "Sloop." The arrow indicates the direction of the slope. A is the azimuth plot at 30 degrees elevation. B shows the elevation plot along the axis of maximum gain, 45-225 degrees.

35 feet away from the center connector on each side. [You may wish to use a ceramic insulator at the side and bottom tie-line attachment points, particularly if high power will be used; see Figure 4.—Ed.] Connect 50-Ω coaxial cable such as RG-8 or RG-58 to the connector and raise the feed point to a height of 30 to 40 feet. Pull the side tie lines sideways and down until the upper half of the antenna forms a taut 90-degree angle and slopes at 30 to 45 degrees with respect to the ground (see Figure 3). Tie off these lines. Attach a short (2-3 foot) length of line to the bottom point of the loop and tie off the bottom of the loop to a stake or a fence post.

The loop will need to be pruned for the antenna to resonate at the desired frequencies. To do this without raising and lowering the antenna for each adjustment, remove lengths of wire at the bottom of the loop and then solder the ends back together. Shorten the loop a few inches at a time until the SWR approaches 1:1 at the desired 40-meter frequency. Adding wire will lower the resonant frequency on all bands.

In my case, a final length of 136 feet yielded SWR values lower than 3:1 over the entire 40, 20 and 15-meter bands. The loop also produced a 2:1 SWR over almost 1 MHz of the 10 meter band (see Figure 1). Since I typically hang out in the phone sections of these bands, my antenna was tuned for the best match there. My old Kenwood TS-830 and ancient Hallicrafters HT-41 kilowatt amplifier—both with adjustable pi matching output networks—easily tune to this antenna at any frequency on all four

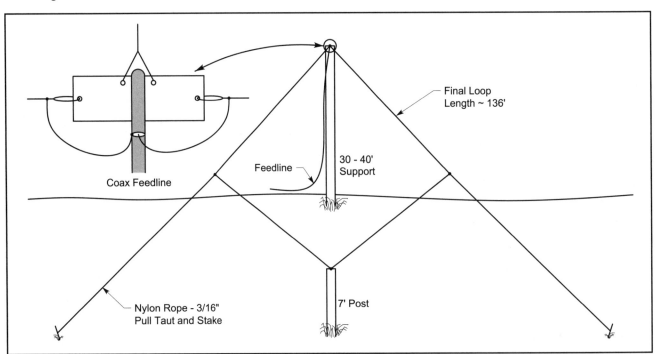

Figure 3—The vertical support of the Single-Masted Sloop can be a mast, tree, building, flagpole, and so on. The simplicity of the design and the multidirectional gain delivered at the harmonics make this antenna a good candidate for Field Day.

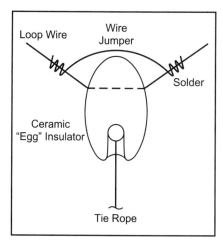

Figure 4—One simple method of attaching tie ropes to wire antennas.

bands. Most recently manufactured rigs can handle the 2 or 3:1 SWR at the band edges. [To lessen the SWR, particularly at higher frequencies, the loop can also be fed with open-wire line.—Ed.]

Results

The results with this antenna are gratifying, especially given that it can be built in a couple of hours from scrap wire and hardware, tunes easily, doesn't need to be elevated to great height and occupies a reasonable "footprint." Stations in Europe, Japan, South America and the Azores were worked with 100 W on 20, 15 and 10 meters within an hour or so of completion and with good signal reports. I tried the antenna on 40 meters during the November 2001 Sweepstakes to get some idea of its performance on that band. I was pleased to find that contacts could be made with the antenna on both coasts from central Ohio at midday in spite of *EZNEC* showing much of the energy on 40 meters radiates straight up (see Figure 2). The performance, simplicity and cost of this antenna suggest to me that this would be the antenna I would roll up and take along on that low-budget DXpedition to the Caribbean.

Notes

[1] Doug DeMaw, W1FB, "A Closer Look at Horizontal Loop Antennas," *QST*, May 1990, p 28.
[2] See Note 1.
[3] *EZNEC* 3.0 Antenna Design Software by Roy Lewallen, W7EL (**www.eznec.com/**; **w7el@eznec.com**).
[4] Chapter 20 ("Antennas and Projects") of any recent *ARRL Handbook* contains drawings that illustrate ways of attaching a center connector.

Rick Rogers, KI8GX, was first licensed as WN-6HGY in 1968, followed by WA6EZT, N9COO and N7GEF. He is a professor of Neuroscience at Louisiana State University where he does research on, and teaches, neurophysiology and physiological instrumentation. A chance encounter with a neighborhood ham cleaning out his garage (SamWestfall, K6PHH) in 1967 started Rick, then 13 years old, on his career in ham radio and science. Ham radio provided an ideal entry point into neurophysiology since the principal elements of the nervous system neurons "talk" to each other using frequency modulated electrical pulses. The ARRL Handbook for Radio Amateurs can be found on the bookshelves of quite a few neuroscientists, whether they are hams or not. You can reach the author at 9831 Bank St, Clinton, LA 70722; **rogersrc@pbrc.edu**.

By Steve Ford, WB8IMY

One Stealthy Delta

This HF antenna keeps a low visual profile while attracting plenty of attention on the air.

Loop antennas have always fascinated me. From a common-sense standpoint they seem impossible. I mean, how can you have a short circuit at the output of your transceiver and call it an antenna? I'd call it bright flash, smoke, and stream of obscenities.

But the magic we call radio is never so straightforward. Yes, a loop antenna is unquestionably a short circuit at the output of your radio—if your radio produced dc. Radio frequency energy, however, is ac and it views a loop quite differently. A loop represents an impedance load to RF. The impedance value depends on the size of the loop, the frequency of the RF and other factors, but it is most definitely *not* a short circuit.

The October 1998 *QST* carried an article of mine titled "One Stealthy Wire" in which I used a remotely tuned antenna coupler to match my radio to a random-wire antenna supported by a lonely maple tree in my back yard. If artists and musicians can go through creative "periods" when their muses suddenly decide to speak in different tongues, so can amateurs. The maple tree is still here and

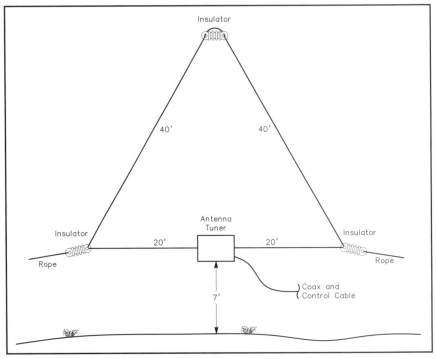

Figure 1—A diagram of the Stealthy Delta.

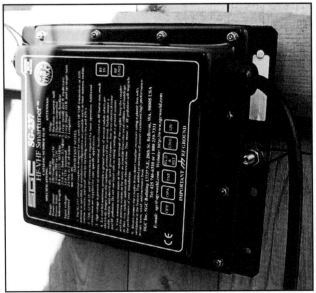

The SG-237 tuner hangs on a wood privacy fence, just behind the tree trunk.

Do you see an antenna in this picture? Probably not!

The privacy fence also acts as an anchor for one of the legs of the Stealthy Delta.

The direct-bury coax makes a discreet jaunt into the bedroom window.

so am I, but I've abandoned my single-wire period and have embarked on the year of the loop. Or to quote Daffy Duck in the memorable cartoon, "I swear, your honor, I will never paint a malicious mustache on a work of art again… *I'm doin' beards now!*"

The Problem Remains the Same

Little else has changed in four years. I still exist on a house lot the size of a postage stamp. The local squirrels have easements written into my deed. I still have a wife who distrusts my every move and despises every antenna I attempt to create. When I wonder aloud about where I can erect my next abomination, her reply is "Cleveland."

I asked Dean Straw, N6BV, our resident ARRL antenna guru, how I could improve my situation. The exchange went something like this…

Dean: Put up a tower and a triband Yagi antenna.

Me: Do these things come with divorce documents?

Dean: How about a 100-foot dipole 50 feet above the ground?

Me: Supported by what two tastefully designed 50-foot objects?

Dean: How about a vertical loop supported by your tree?

That's when the sweatsock-filled-with-nickels-of-inspiration struck me upside the head. How about a loop not only supported by the tree, *but in the tree?*

The Stealthy Delta

A delta loop gets its classy moniker from the Greek alphabet, namely the letter *delta*, or Δ. My Stealthy Delta is a vertical wire triangle fed directly in the middle of its base (see Figure 1). For multiband HF operation the idea is to make the triangle as big as possible. It also helps to keep the base of the triangle about 7 feet or so off the ground.

For my application each side of the triangle is 40 feet in length. Remember that the wire is continuous; that's why they call it a loop. Using our wood privacy fence to hide the bottom wire, I strung the loop out 20 feet to an insulator, up into the tree (to an insulator suspended by a Nylon rope), down to another insulator on the other side of the tree and then finally back to where I began. Was it a perfect triangle? No. Was it good enough for Amateur Radio and rock n' roll? You bet.

And now that it was strung, how would I feed the Stealthy Delta? I would need an antenna tuner for multiband operation—that much was clear. With the tuner indoors I could use 450-Ω ladder line between the tuner and the feed point of the antenna. In my case, however, the ladder line would have to take a torturous route to reach the Stealthy Delta. It would have to careen through the air and directly over my wife's sacred hedges and rose bushes. That was unacceptable (to her, anyway). The alternative was to use a substantial length of buried coax, but coaxial cable is much too lossy in the face of the high SWRs that would exist between the antenna and the tuner.

If the mountain will not come to Mohamet, Mohamet must go to the mountain. Or putting it in a ham context, if the antenna will not come to the tuner without unacceptable feed line loss, the tuner must go to the antenna. Borrowing an idea from my "One Stealthy Wire" article, I invested in a new SG-237 remote automatic antenna tuner from SGC Inc (**www.sgcworld.com**). I installed the tuner at the feed point, hiding it behind the tree trunk, and ran direct-bury coax and a power cable all the way back to the station. I buried most of the wires, except for a short run up the side of the house and into the guest bedroom window.

How Does it Work?

The SG-237 is RF activated. You transmit and it finds a low SWR within a few seconds. That low SWR is achieved at the antenna. With the good-quality coax I used between the tuner and my radio, feed line loss was kept to a minimum (a little over 1 dB on 6 meters and much less on lower bands). With my Stealthy Delta the SGC tuner can find an acceptable match with an SWR less than 2:1 on any amateur frequency from 80 through 6 meters. If I had erected a somewhat larger loop, I probably could have operated the antenna on 160 meters as well.

In terms of performance, the Stealthy Delta is definitely superior to my single stealthy wire. Even on 80 meters, where it is way too short, the loop surprised me. During a recent RTTY contest I made several contacts into Europe on 80 meters, which I've never done before on RTTY from home. On 40 through 10 meters I consistently receive strong signal reports. I worked the XRØX and TI9M DXpeditions on RTTY after just a few calls and even managed to get through the pileup to work the PWØT group on 15-meter SSTV. Not bad for a wire triangle.

And best of all, the Stealthy Delta is very stealthy indeed. The tree camouflages most of the antenna. The photos that accompany this article were shot in March when the tree was bare and yet the antenna is very difficult to see. Just imagine how invisible it is when the tree is in bloom.

Will I stick with the Stealthy Delta? Certainly...for now. I can't beat the performance and convenience, but I'm sure I'll eventually think of something that will. Some day my "loop period" will give way to some other source of annoyance for my wife and child…

"I swear, honey, I will never erect another diabolical delta…*I'm doin' rhombics now!*"

You can contact the author at ARRL Headquarters, 225 Main St, Newington, CT 06111; **sford@arrl.org**.

By John S. Belrose, VE2CV

A Horizontal Loop for 80-Meter DX

Working DX on 80 meters doesn't necessarily require big towers or trees. This 80-meter quad loop system requires only supports of modest height—a better single-element antenna may be hard to find.

Introduction

In 1997 the author published an article on vertical full (and ground plane type half) wave loops for 80 meter DX.[1] In that article it was noted that perpendicular (horizontal) polarization is the preferred polarization, particularly at low elevation angles, since horizontally polarized waves are hardly affected by the finite conductivity of the ground in front of the antenna. An exception when vertical antennas come into their own is a vertically polarized antenna over very good ground, near the seashore or over alkaline salt flats.

A practical 80 meter horizontal dipole is, however, not an ideal antenna for DX. For optimum communications with distant stations the antenna's radiation pattern should have a null overhead, to minimize near vertical incidence sky-wave signals from atmospheric noise and interference, and a low angle lobe to maximize reception/transmission over paths to distant stations. To achieve such a pattern with a half-wave dipole it would be necessary to install the dipole a half wavelength above the ground, that is to say, at a height of 40 meters for the 80 meter band. This is impractical in many instances.

A full-wave horizontal loop for the 80 meter band at a practical height of 15 meters is a popular antenna nicknamed a "Loop Skywire" that has been in *The ARRL Antenna Book* for years. In the author's view this antenna does not have the desired radiation pattern for 80 meter DX. Aside from the fact that the direct and ground-reflected waves reinforce at an elevation angle of 90°, the loop itself has some directivity in this broadside direction. Doug DeMaw has referred to such an antenna as a "cloud warmer."

Paul Carr, N4PC proposed a solution for this problem.[2] He fed diagonally opposite corners of a square loop with equal but oppositely phased currents. For a full-wave loop this produces a null in the overhead radiation pattern, akin to the time-honored W8JK array—a pair of closely spaced dipoles fed out of phase.[3] For 80 meters N4PC used a $^3/_4$-wavelength loop, which had the desired elevation pattern, but with corner feed the azimuthal pattern is skewed compared with the loop modeled by the author.

According to the author's simulation using W7EL's *EZNEC Pro* version of the numerical electromagnetic code NEC-4D, the antenna's impedance at the input to a transmission line feeder of practical length, is not a particularly convenient value to tune and match. N4PC did not comment on this. In fact, he did not include the phasing and the feeder transmission lines in his model, and so he could not comment on the input impedance of his antenna system although he reported "no problem with tuning and matching his antenna on all bands 80 through 10 meters."

In this article the author uses numerical simulation to address the radiation characteristics and the tuning and matching details of a symmetrical full wave quad loop designed specifically for 80 meter DX.

A Horizontal Square Loop with a W8JK-Like Radiation Pattern

W. Bolt, DJ4VM, described a multiband

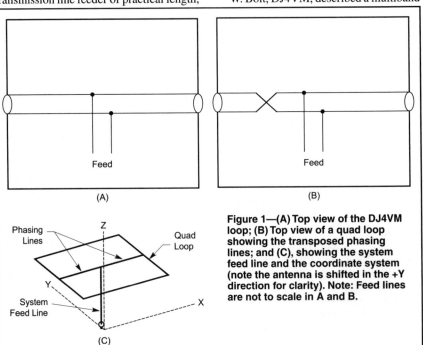

Figure 1—(A) Top view of the DJ4VM loop; (B) Top view of a quad loop showing the transposed phasing lines; and (C), showing the system feed line and the coordinate system (note the antenna is shifted in the +Y direction for clarity). Note: Feed lines are not to scale in A and B.

[1]Notes appear on page 67.

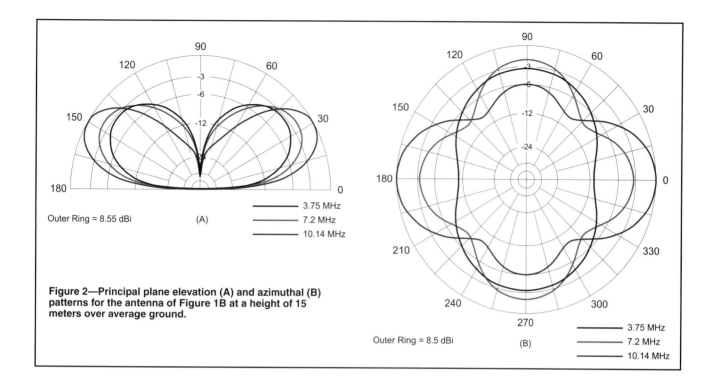

Figure 2—Principal plane elevation (A) and azimuthal (B) patterns for the antenna of Figure 1B at a height of 15 meters over average ground.

vertical quad loop with both of the vertical sides fed in-phase by means of a "phasing line."[4] This symmetrical feed arrangement (Figure 1A) has the advantage of ensuring a symmetrical current distribution on the loop—and hence a "clean" radiation pattern over several bands (40 meters to 10 meters).

Our interest here, however, is a horizontally polarized loop operating in "W8JK mode." Rotating the plane of his loop 90°, we now have the horizontal loop, in which the system feed line connecting at the center of the phasing lines can drop vertically at right angles to the plane of the loop (see Figure 1C). The symmetry of this loop arrangement is appealing.

The author decided to carry out a detailed numerical modeling study for this loop fed with a balanced, transposed phasing transmission line (opposite sides of the loop fed out of phase). The horizontal full wave loop, λ/4 or 20 meters on a side for 3.75 MHz, at a height of 15 meters over average ground, is numerically modeled as three separate cases:

Case 1) Out-of-phase sources placed at the centers of the sides of the loop (the sides parallel to the Y-Z plane);

Case 2) Wires are added for the conductors of a 600-Ω phasing line between the opposite sides with the source placed at the middle. (Note: The conductors for the transmission line feeding the side of the loop in the –X direction are transposed to provide the out-of-phase feed; see Figure 1B.)

Case 3) Case 3 differs from Case 2 only in that a system feed transmission line is added, with the source on a jumper wire between the transmission line conductors at its bottom or transmitter end (Figure 1C).

All wires are no. 12 copper. The spacing for the conductors of the transmission lines is 150 mm (for an impedance $Z_o = 600$ Ω). The length of the transmission line feeder is 14 meters.

The reason for the three-case modeling sequence is to be sure that the transmission lines included in the model are performing correctly. We find that they are by computing the feedpoint impedances of the two models with different techniques. The *NEC-4D* input impedance for the Case 3 antenna at 3.75 MHz is $9.8 + j462$ Ω. For the Case 2 antenna, fed by a 600-Ω open wire transmission line 14 meters long, the input impedance is $9.9 + j464$ Ω according to *TL*, the transmission line program by N6BV and published by the ARRL.

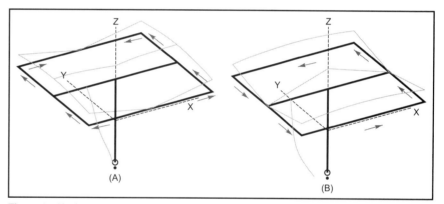

Figure 3—Horizontal quad loops showing currents on the loop wires for the case where the phasing lines provide in-phase feed (A); and out-of-phase feed (B). Note the left phasing line is the transposed feeder. The frequency is 3.75 MHz.

While the source impedances are, of course, different for all three configurations, radiation patterns are substantially identical. The maximum gain is decreased slightly with the addition of the transmission lines as would be expected because of the high SWR on the transmission lines (see below). At 3.75 MHz, the gains in the principal plane (the Y-Z plane) are 6.02 dBi, 5.34 dBi and 3.93 dBi, for cases 1, 2 and 3, respectively. From this point, all discussion will be of the antenna system of Case 3 and "system impedance" will refer to the impedance at the transmitter end of the common feed line.

The principal plane radiation pattern is shown in Figure 2 at a frequency of 3.75 MHz. For interest, the pattern is also shown at 7.2 MHz, and 10.14 MHz. The calculated antenna system impedances and radiation characteristics (gain and take-off angle, ψ, are given in Table 1).

We are concerned here with the radiation pattern for 80 meter DX. Clearly the radiation patterns shown in Figure 2 are almost ideal: a deep overhead null, and a bidirectional pattern with a take-off angle (41°) that is low for the practical height (15 m) of the loop. The azimuthal pattern is also good (for an 80 meter antenna) with a front-to-side ratio about 10 dB.

Clearly, at 3.75 MHz, conductor loss in the system feeder transmission line attached to the junction of the phasing lines will be an important consideration because of the high SWR. For an open wire line 14 meters long made of no.12 wire the transmission line loss is 1.5 dB, according to NEC-4D. By using a larger diameter wire, the loss can be reduced. As an example of the benefits of using heavy-duty transmission lines, if the conductors were no. 4 copper wire, the transmission line loss would be 0.5 dB. For this calculation the spacing for the larger diameter wire is 150 mm so that Z_o is less than 600 Ω, but the characteristic impedance for this feeder is relatively unimportant. The transmission line loss at 7.2 MHz and 10.14 MHz is negligible. [While quite large, no. 4 wire is commonly used for electrical grounding. The outer shield of coaxial cable could also be used to construct an open wire line with large conductors.—Ed.]

Table 1

Impedance at the Junction of the Phasing Lines for Out of Phase Feed (W8JK-like Mode)

Frequency (MHz)	Impedance (Ω)	Gain (dBi)	Take-off Angle (ψ)
3.75	5.5 – j 310	5.34	41°
7.2	285 + j 654	6.61	34°
10.14	73 + j 84	8.55	32°

Current Distribution on the Loop

To understand how the patterns for in-phase and out-of-phase feed come to be, it is interesting to look at the current distributions on the loop. Figure 3 shows the current distribution (amplitude only) and by arrows the relative phase relationship for the DJ4VM-type loop with opposite sides fed in phase, and for the same loop configured with opposite sides of the loop fed out-of-phase to radiate in "W8JK mode." Phase information is useful in determining that certain kinds of antennas are modeled correctly. For closed loops, particularly for the case with dual sources, the phase as calculated by NEC can be confusing.

Sudden reversals in phase may not be a cause of concern if they result from the way the wires have been defined. Positive current is defined as being from end 1 to end 2 of the wire, so if for two wires end 1 is connected to the other end 1, a 180° shift in current phase will be indicated at the junction of the wires. The actual current is continuous, as it should be, but the direction reference changes from one wire to the other. This can lead to confusion, particularly for closed loops such as our present model where in Case 1 the loop is fed in the center of two sides (two sources and no transmission lines). If the quad loop is modeled with end 1 connecting to end 2 for all wires, opposite sides will have wires defined in opposing directions. For the desired radiation pattern, the phase relationship of the sources in the model will then be 180° different from those of the actual antenna.

That is why in Figure 3 we show the relative phase relationships sorted out by the author by arrows. In the Case 1 model with two sources the wires have been defined so that with source current phases of 0° and 180° the pattern is as expected. Figure 3B shows clearly that the phases of currents in the wires parallel to the X-Z plane are out of phase (the desired W8JK mode). For in-phase feed (the DJ4VM antenna) of Figure 3A, the currents

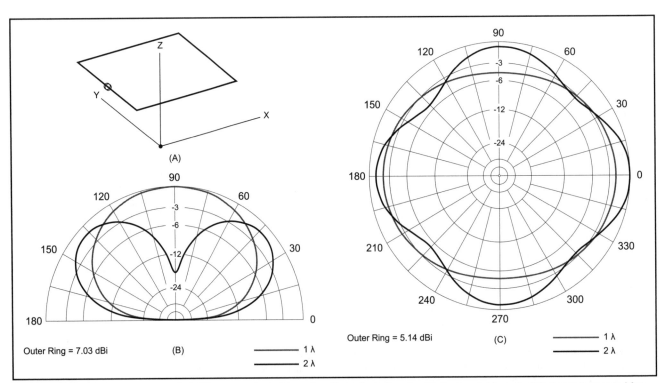

Figure 4—Elevation (B) and azimuthal (C) patterns for a 1-wavelength and a 2-wavelength horizontal loop (a 160 meter loop used for 80 meters). Note that for the 1-wavelength loop the azimuthal pattern for an elevation angle of 45° is plotted since the take-off angle is 90°.

Matching Network Capacitor Values

Let R1 = 6.8 Ω and R2 = 50 Ω (for a 50-Ω coaxial feed).

The series inductor $X_L = [R1 (R2 - R1)]^{1/2} = [6.8 (50 - 6.8)]^{1/2} = 17$ Ω

Because the system presents an inductive reactance of 343 Ω, cancel all but 17 Ω with a series capacitor of (343 – 17) = 326 Ω (130 pF at 4 kV rating). The impedance match is very sensitive to this reactance, so a variable capacitor is required.

The capacitor to ground $X_c = R2 [R1/ (R2 - R1)]^{1/2} = 50 [6.8/(50 - 6.8)]^{1/2} = 19.8$ Ω (2139 pF). This is a low voltage capacitor—approx 300 V for 1500 W at 50 Ω—and may be of fixed value.

Table 2
System Impedance

Frequency (MHz)	Impedance Ω
3.75	6.8 + j343
7.2	135 – j 37
10.14	67 – j12

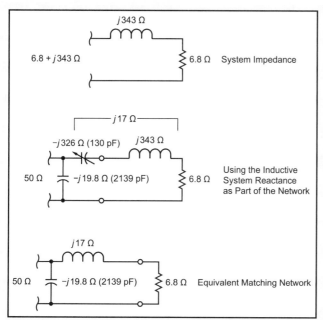

Figure A—The three steps of designing the "step-up" L-match for an inductive load.

on the wires parallel to the Y-Z plane are in phase. With this arrangement we have a "cloud warmer" antenna.

A Practical Installation

For a practical installation of the loop the system impedance is an important consideration (as for any single band or multiband antenna system that depends on tuning feed line impedances). The impedances to match are given in Table 2 assuming the heavy-duty low loss transmission line (no. 4 wire spaced 150 mm, length 14 m).

A balanced Antenna System Tuning Unit (ASTU), or an unbalanced ASTU with a "common subchassis" isolated internal ground, as described in the *2002 ARRL Handbook* (pp 22.56 ff), could be used, with a 1:1 current balun between the ASTU and the transmitter. The author describes a "special ASTU" for the antenna below. [ASTU is used as opposed to the more common ATU because the tuning performed is of the complete antenna system including the feed line, as opposed to just tuning the antenna.—*Ed.*]

We are concerned with keeping transmission lines losses low for the loop when operated on the 80 meter band (note the high inductive reactance compared with the resistive component in Table 2). Let us now consider losses in the ASTU. Power loss in ASTUs can also be an important consideration, but in this case, the power loss in the tuner can be minimized if you fabricate a "special ASTU" for this band.

The inductive component of the system reactance could be cancelled with a series capacitor, leaving a 6.8 Ω resistive impedance. The simplest network to match the low, resistive impedance to a higher value (6.8) is the L-match network, comprising a series inductor and on the transmitter side a capacitor to ground. But we do not need the series inductor because the feedpoint impedance is already inductive. Hence what we need is a series capacitor to cancel all but the necessary matching inductance and the capacitor to ground. The author described this tuning arrangement in 1953,[5] for matching center-loaded mobile whips. Great—we can build a simple and efficient ASTU for our 80 meter quad loop!

A Comparison with a "Sky Wire" Loop

Paul Reed, VE2LR, who intends to put up a 160 meter horizontal loop, brought to the author's attention an article by Richard Stroud, W9SR,[6] who has erected a 160 meter full wave horizontal or "sky wire" loop fed at a single point. He claims that his loop has "opened up a new world of DX-ing." The reader of the present article may well say, "For 80 meter DX is it worthwhile feeding the opposite sides of a loop out of phase, since I can more easily put up a skywire loop?"

The 2-wavelength perimeter loop does produce a null overhead—see Figure 4. The "skywire loop" has a gain of 5.1 dBi (take-off angle 47°) at a frequency of 3.75 MHz—compared with 4.8 dBi (take-off angle 41°) for our phased loop with a heavy-duty low loss feeder (3.8 dBi if the no. 12 feeder wires are used).

However, comparing the horizontal pattern of the 2 λ skywire loop to the out-of-phase 80 meter loop in Figure 2, the latter has a nice, clean directional pattern. For the DXer who wants to work stations in a preferred direction, 4.8 dBi directive gain with a front/side ratio of about 10 dB is an attractive antenna and the 6° lower take-off angle is well worth having. Try the loop and you will see! Increasing transmitter power can to some extent replace antenna gain, but you can only work stations that you can hear.

Comparing Vertical and Horizontal Loops

At the outset it was noted that if the ground conductivity in front of the antenna was poor, horizontal was the preferred polarization. Since the author's earlier article (Note 1) extolled the performance of vertical delta and quad loops, let us compare the performance of the horizontal quad loop with out-of-phase feed to that of a full wave delta loop on a support of the same height (15 m), for two ground conductivities: average ground (σ = 5 mS/m, ε = 13) and poor ground (σ = 1 mS/m, ε = 3). As can be clearly seen in Figure 5, for poor ground the horizontal loop wins hands down. For average ground the

65

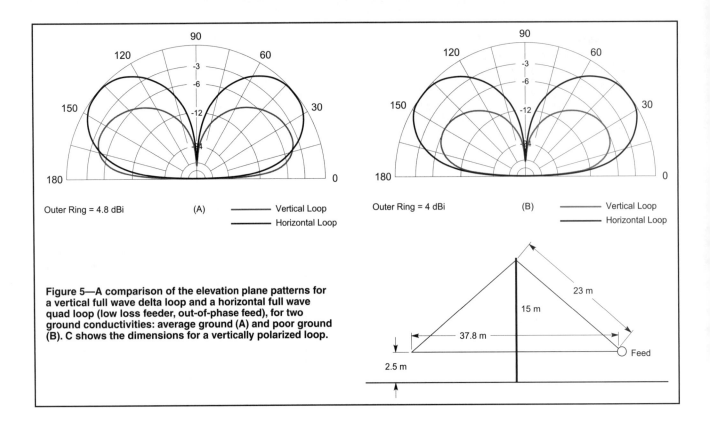

Figure 5—A comparison of the elevation plane patterns for a vertical full wave delta loop and a horizontal full wave quad loop (low loss feeder, out-of-phase feed), for two ground conductivities: average ground (A) and poor ground (B). C shows the dimensions for a vertically polarized loop.

vertical delta loop outperforms the horizontal quad loop only for signals arriving at very low elevation angles (less than 15°).

Another consideration is background noise. The author has no side-by-side comparison of background noise for the two loop systems, vertical and horizontal, only anecdotal evidence that the noise level on a horizontal loop would be lower. Those who have used horizontal loops often report that "the residual noise level is very low," and "very seldom is the band completely dead."

A Note about Bandwidth

The antenna system bandwidth is narrow, 16 kHz (at 3.75 MHz) for a 2:1 mismatch. The operational bandwidth (estimated assuming a conjugate match) would be about 26 kHz based on the author's experience. Clearly the series capacitor should be remotely controlled.

Conclusions

While the horizontal loop antenna with opposite sides fed out of phase could be an excellent single-element for 80 meter DX, it is not recommended that it be the only 80 meter antenna for station use. An ordinary dipole is traditionally considered to be a necessary part of the antenna complement for operations on the 80 meter band, since a pattern optimized for high angle sky-wave, for short to medium range paths, is required for normal operation on this band.

The horizontal loop described can, however, be arranged to provide both low and high angle radiation patterns, with additional complication. If both sides of the loop are fed in phase, then we have the desired pattern for the short to medium distance range. If both sides of the loop are fed out of phase we have the desired pattern for DX. A relay located at the center of the antenna system to switch the phase of one of the phasing lines is required to facilitate pattern selection.

This would require five supports—one for each corner and a center pole to support the phasing lines and the heavy-duty feed line. [A method to avoid the center pole would be to bring the two phasing lines directly to the ground-level system feedpoint at the center or to drop them vertically from the loop wires and then run them parallel to the ground to the center point. Doing so would alter the system feed point impedance and require changes to the matching network.—Ed.]

The ability to change patterns with the same antenna is attractive, but this requires two ASTUs because the antenna system impedance for in-phase feed is very different as shown in Table 3 (compare with Table 2).

Figure 6 shows how a remote "special ASTU" should be used for the 80 meter out-of-phase fed loop, with a coaxial cable transmission line to the "shack"; and the open wire line could be brought into the "shack" for the higher bands and for the in-phase fed loop. As an alternative, different lengths of feed line for the 80 meter in-phase fed loop could make tuning and matching more practical. This would require another transmission line relay.

Finally, if you have the room and wish to use the design for both 160 meter and 80 meter DX, make the loop twice as big (40 m on a side). This shifts concerns about keeping transmission line and ASTU losses low to the 160 meter band. The 160 meter radiation patterns will be similar to those computed for 80 meters for our "half size" loop, and the patterns for 80 meters will be similar to those computed for 40 meters.

Acknowledgments

The author works (part time) for the Communications Research Centre Canada, Ottawa, Ontario; hence computing facilities not available to the average amateur in radio are available. In particular, the author is licensed to use the *NEC-4* program.

Table 3
System Impedance Fed In-Phase

Frequency (MHz)	Impedance (Ω)
3.5	383 + j1794
3.75	5515 + j1904
4.0	863 − j1595
7.2	102 − j1094
10.14	202 + j774

Notes

[1] Belrose, J.S., "Loops for 80m DX," *QEX*, Aug 1997, pp 3-16.
[2] P. Carr, "The N4PC Loop," *CQ* Magazine, Dec 1990, pp 11-15.
[3] J. Kraus, "The W8JK Antenna: Recap and Update," *QST*, Jun 1982, pp 11-14.
[4] W. Boldt, "A New Multi-band Quad Antenna," *Ham Radio Magazine*, Aug 1969; see also L. A. Moxon, *HF Antennas for all Locations*, published by The Radio Society of Great Britain, 1982, pp 158 and 160.
[5] Belrose, J.S., "Short Antennas for Mobile Operation," *QST*, Sep 1953, pp 30-35, 108.
[6] Stroud, R.W., "A Large, Remote-Tuned Loop for HF DX," *CQ* Magazine, Jul 2001, pp 44-54.

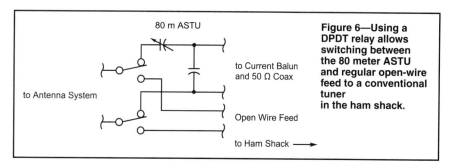

Figure 6—Using a DPDT relay allows switching between the 80 meter ASTU and regular open-wire feed to a conventional tuner in the ham shack.

John S. (Jack) Belrose received his BASc and MASc degrees in Electrical Engineering from the University of British Columbia, Vancouver, in 1950 and 1952. He joined the Radio Propagation Laboratory of the Defence Research Board, Ottawa, Ontario, in September 1951. In 1953 he was awarded an Athlone Fellowship, was accepted by St John's College, Cambridge, England and by the Cambridge University as a PhD candidate, to study with the late Mr J. A. Ratcliffe, then Head of the Radio Group, Cavendish Laboratories. He received his PhD degree from the University of Cambridge (PhD Cantab) in Radio Physics in 1958. From 1957 to present he has been with the Communications Research Centre (formerly Defence Research Telecommunications Establishment), where until recently (19 December 1998) he was Director of the Radio Sciences Branch. Currently he is working (part time) at CRC (2 days/week) devoting his time to radioscience research in the fields of antennas and propagation—a sort of transition to full retirement.

Dr Belrose was Deputy and then Chairman of the AGARD (Advisory Group for Aerospace Research and Development) Electromagnetic Propagation Panel from 1979-1983. He was a Special Rapporteur for ITU-Radiocommunication Study Group 3 concerned with LF and VLF Propagation. He is an ARRL Technical Advisor in the areas of radio communications technology, antennas and propagation; a Fellow member of the Radio Club of America and Life Senior Member of the IEEE (AP-S). He has been a licensed radio amateur since 1947 (present call sign VE2CV). You can reach the author at 17 rue de Tadoussac, Aylmer, QC J9J 1G1, Canada; **john. belrose@crc.ca**.

Tips for Field Day Power

Power Packing for Emergencies

John S. Raydo, KØIZ

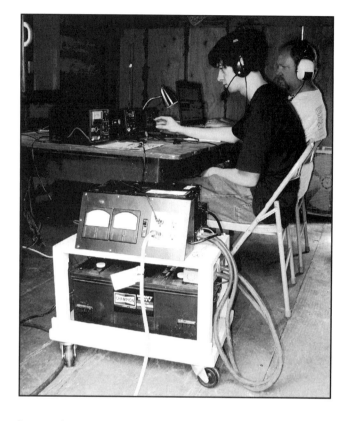

Recent natural disasters have reemphasized for me the need for emergency power. Portable generators work well but can be highly inefficient under low loads. So how could I make a supply of gasoline last longer? A heavy-duty uninterruptible power supply (UPS) was the answer.

I was familiar with the small UPS units sold for use with PCs. These are typically rated for 500 VA or so and contain a small gel-cell battery. While adequate to power a PC for a few minutes the capacity is not nearly enough for reasonable emergency use. I might need to power ham gear, some lights, a PC, perhaps a small TV and other gear. More power was needed.

I decided I could make my own custom UPS from a battery charger, large capacity battery, dc to ac power inverter and control, indicator and distribution circuitry. My UPS requirements began to firm up:

- 1000 W continuous power available if needed.
- Sufficient capacity to last for X hours.
- Compact and portable configuration.
- Total cost target — under $600.

Picking an Inverter

I had first considered a modified sine wave inverter. These are commonly available in sizes from 100 W and up at very reasonable prices. The larger ones would have adequate capacity. However, I had some concerns in powering expensive ham gear with anything other than an ac sine wave. Others have also reported RFI from the modified sine wave type inverters.

I finally selected a Tripp-Lite Model APS1012 inverter. It had a 1000 W rating and many other features including my preference for true sine wave output. Key benefits included auto switchover to battery, a sophisticated three-step internal battery charger, tightly regulated frequency, over/under adjustable voltage limits and numerous other selectable settings. I have not experienced any RF noise on the HF bands with the Tripp-Lite unit. It is also dependable. My business used one for telephone and server backup with 40,000 hours of continuous operation.

This model has since been replaced by the APS1250, which is smaller and even higher rated at a continuous 1250 W output.[1] It is available from several Internet sources for about $400. There are also other brands with similar features.

A block diagram of the resulting system is shown in Figure 1.

Picking a Battery

The most familiar battery is a "flooded" (wet) type such as used in most cars. Some have removable caps to allow addition of

[1]Notes appear on page 71.

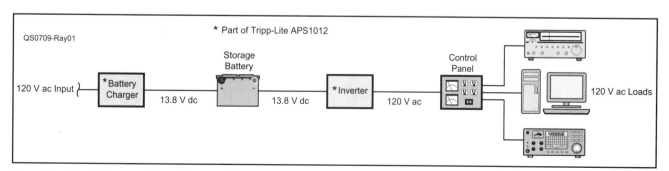

Figure 1 — Block diagram of the uninterruptible power system.

water. The so-called "maintenance free" variety is not totally sealed, allowing hydrogen gas to vent during charging. To avoid the danger of explosion, ventilation is required for indoor use. Your auto battery is designed to supply a very large starting current for a very short time. It contains many thin plates that look like a fine foam sponge designed to provide maximum surface area. Starting batteries are usually rated in CCA (cold cranking amperes) rating rather than ampere-hours (Ah). A relatively small number of deep discharge and charge cycles will cause this type of battery to fail. Thus a starting battery is not a good choice for emergency power.

Deep cycle flooded batteries have much thicker solid plates and can be repeatedly discharged as much as 80%. They provide high capacity at lowest cost with a typical life of 5 to 8 years. The most common are golf-cart GC-2 batteries. Don't confuse deep cycle with "marine" batteries. Many marine batteries are a cross between starting and deep cycle construction. An MCA rating (marine cranking amperes) probably means it's not a true deep cycle and likely will not last long in this application.

A "gelled" battery type contains acid with added silica gel to turn the acid into something that looks like gelatin dessert. This means no spill. Some, however, have to be charged more slowly than either flooded lead acid or absorbed glass mat (AGM) (see below) batteries to avoid gas pockets forming on the plates and forcing the gelled electrolyte away from the plates. For an emergency power pack with fast recharge rates this would be a disadvantage.

A newer type of deep cycle battery is the AGM. A fine boron silicate glass mat saturated with acid is placed between battery plates. AGM batteries have many advantages including a battery life of up to 15 years. AGM batteries generally cost considerably more than the other technologies, however. Unlike flooded batteries the oxygen and hydrogen produced in an AGM battery during charging are 99% recombined. This means almost no water is lost. AGM batteries will not spill and will lose only a few percent of charge per month. This means longer periods between charging than flooded batteries.

Next, how big a battery do we need? The battery Ah rating is usually specified at a 20 hour discharge rate (12.5 A for 20 h equals a 250 Ah rating). Since all batteries have internal resistance the Ah rating will be lower at higher discharge rates. At an 8 hour discharge rate, most batteries will supply only about 80% of rated capacity and substantially less at higher discharge rates. Thus, for a high power emergency supply it might be best to reduce the Ah rating by 20 or 30%.

The depth of discharge affects another important factor — battery life. Discharging only to 50% will typically yield 1000 discharge-charge cycles, double that of 80% discharges. Forget starter type batteries — they will fail after only a couple of dozen cycles. Complete discharge will badly damage most all batteries. You may have observed this effect if you have ever left your car lights on overnight. The best strategy here is to buy a big battery with lots of capacity so depth of discharge will be modest.

Other than initial cost I think the best choice is a big AGM battery. However, I went the cheap route with a big brute (135 pound) 8D size industrial flooded battery available at discount clubs for about $125. So far it's doing fine. Another low-cost option would be two GC-2 6-V batteries connected in series.

Assembling the Power Pack

The battery sits on a 15 × 25 inch piece of ¾ inch plywood with small ¾ × ¾ inch wood strips nailed on top at the edges. The inverter and control box are mounted on a 13½ × 23 inch piece of ¾ inch plywood with 1½ × 1½ inch wooden legs 12 inches high. The inverter assembly is not attached to the battery cart and can be lifted off for battery maintenance or travel. Four heavy duty 3 inch casters and some paint complete the UPS cart.

The control box is optional since loads can be plugged directly into the inverter ac outlets. However, I found it very handy to monitor the ac input voltage (especially if from a generator), and the UPS output voltage and load.

My control box includes two meters from my junk box, two duplex outlets, a circuit breaker and a neon voltage indicator. Almost any type of dc or ac meters can be adapted for this application. My junk box voltmeter displays an expanded 100 to 140 V scale based on a circuit in an earlier *QST* Technical Correspondence.[2] A 150 V ac meter or a dc meter with a 1N4004 diode and appropriate series resistor should also work. Figure 2 is a schematic of my control box.

My "wattmeter" actually measures ac

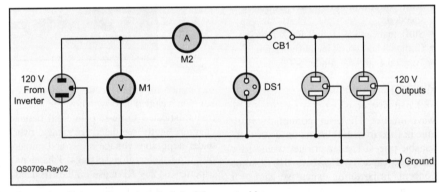

Figure 2 — Schematic and parts list of the control box.

CB1 — 10 A circuit breaker, Grainger 5B738 or equivalent load center breaker. Use hot glue to fasten to control panel.

DS1 — 120 V neon pilot indicator with internal dropping resistor.
M1 — 150 V ac meter (see text).
M2 — 10 A ac meter or wattmeter (see text).

Figure 3 — Control box showing voltage meter, "wattmeter," circuit breaker and ac outlets.

Figure 4 — Battery positive terminal. The 200 A fuse is on left. The link from battery terminal to fuse is made from a 2½ inch length of flattened ½ inch copper water pipe.

amperes with the "watts" calibration based upon an assumed 120 V. Accuracy seems adequate for the purpose but a true ac wattmeter would be better. Meter scales were replaced with new scales that I produced using a nice little program from WB6BLD.[3]

The control box itself is made of scrap ¾ inch wood with ¼ inch wood front and back. Dimensions are 7 × 15½ × 4¾ inches (HWD), sloping to 1¾ inches deep at the top, to allow better visibility of the meters. There is nothing unique about these dimensions or type of construction. A reclaimed computer ac cord was cut to about 2 foot length and plugs into one of the inverter ac outlets. Figure 3 shows the completed control box.

The inverter is connected to the battery using two 3 foot lengths of 00 gauge welding cable available from welding supply stores. It is much more flexible than other types of cables of this size. Stripping the insulation off the ends requires special attention. If provisions are not made, the many fine strands tend to flare out making connection at the inverter quite difficult. I soldered the wire strands together at the ends using a propane torch and then used a file to trim to fit into the inverter terminals.

Each of the battery terminals is shielded from accidental shorts with a standard 4 × 4 inch plastic electrical box with a hole drilled for the battery terminal. The box on the positive terminal also contains a 200 A fuse (see Figure 4).

Be Safe

These big batteries pack lots of energy. Make sure you install a fuse at the battery. Mark the positive terminal box and cable with red paint or red electrical tape. Take particular care when using a wrench to tighten those terminals. A shorted battery can explode!

AGM (and gel) batteries are safer to handle than flooded batteries, since they are sealed and don't have free liquid. If you use a flooded battery, some additional safety rules should be followed:
- Never add acid except to replace spilled liquid.
- Only use distilled water to top off non-sealed batteries.
- Charging non-sealed batteries will cause venting of explosive hydrogen gas. Keep away from flames including your furnace and have good ventilation.
- Overcharging will damage batteries and cause excessive venting and even splattering of corrosive liquid.
- Sulfuric acid will burn skin and destroy clothing. If any is spilled wash it away with a large quantity of water. Protect your eyes. Fumes are also very dangerous and should not be inhaled.
- All big batteries are heavy — watch your back!

Performance

My 8D battery contains 2400 watt-hours (250 Ah × 12 V × 80%) of energy to the 80% discharge point. The inverter is rated at 90% efficiency. Thus, this emergency power pack could provide 100 W of clean continuous ac power for 21 hours, 300 W for about 7 hours, and even 1000 W for a while. The low duty cycle of SSB and CW operation means I'll have the capability to support extended emergency operations. More expensive 8D batteries have even greater capacity.

A noisy ac generator is a necessary evil for most ARRL Field Day operations. At our radio club's last three Field Days we connected my power pack between the generator and our radios. Our three HF/VHF radios ran the majority of time on the power pack with only intermittent use of the generator to operate and recharge. Less gasoline was used and less noise was produced.

Power outages at our home have provided another good use of the power pack. I normally keep the Power Pack plugged into a 120 V ac outlet so the internal battery charger keeps the battery topped off without overcharging. During a recent power outage my wife, Judy, connected our big TV, satellite receiver and several lights to the power pack using an extension.

Notes
[1]**www.tripplite.com/products/**.
[2]J. Grebenkemper, KI6WX, "An Inexpensive, Expanded-Range Analog Voltmeter," *QST*, Dec 1992, p 52, and "Expanded-range DC and AC Voltmeters," Technical Correspondence, *QST*, May 1993, p 77.
[3]W. Jones, K8CU, "Easy Custom Meter Faces at Home," Hints & Kinks, *QST*, Oct 2002, pp 63-64. Also see **www.tonnesoftware.com**.

John Raydo, KØIZ, received his Novice class license in 1957 at the age of 13. From early on he has enjoyed designing and building ham radio equipment and antennas. He has authored several previous articles for QST. *He is an active member of the Johnson County (Kansas) Radio Amateur Club. John is a graduate electrical engineer and also holds a liberal arts degree in math and science as well as an MBA. He started his career working in the engineering department of TWA and later headed up their information services and purchasing departments. He is now retired from his second career as a principal with an investment company. John can be contacted at* **kcflyers@yahoo.com**.

Modern Portable Power Generators — Small, Sleek and Super Stable!

Thanks to a new breed of ultraquiet portable power generators, going "off the grid" has never been easier.

Kirk A. Kleinschmidt, NTØZ

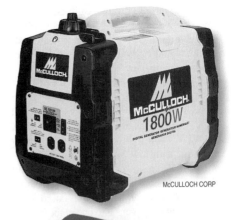
McCULLOCH CORP

AMERICAN HONDA POWER EQUIPMENT

YAMAHA MOTOR CORP USA

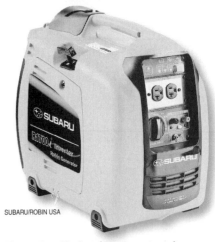
SUBARU/ROBIN USA

Field Day, our annual pilgrimage to the woods, the lake or the school softball field, encompasses a huge variety of operating styles and overall scope. They range from a solo ham with a battery pack and a low power transceiver operating from a secluded mountain lake to a 20 station traveling circus complete with a big top tent, a half dozen towers and three guys handling the PR (and if you look closely, you may even see a clown or two!).

Power Up for Field Day

Every Field Day station needs power, and while there is an operating category for home stations, most use a source other than the ac mains. This makes Field Day a good exercise in preparation for a real disaster in which power is often unavailable. A single QRP transceiver might operate from a small battery pack for several days, while a 100 W rig might last for a similar period while powered from a large deep cycle battery. But at some point, field operations of any size and duration will need power from another source such as a solar panel, a car alternator, a wind turbine or a portable power generator.

If you haven't been paying close attention — as I hadn't — you might not know how far portable generators have come in the past decade.[1] Amazingly, most of the sticking points that accompanied the generators of a decade ago have now been largely overcome — size, weight, noise and, most importantly, voltage and frequency stability.

I also realized that the evolution of portable power generators (*gensets* for short) over the past decade has paralleled the evolution of ham radio gear to a large extent. Over the past 10 to 15 years, ham radio transceivers have become tiny, powerful, portable, reliable and inexpensive. A station that filled a desktop in 1980 is now smaller than a 1980 CB radio — and infinitely more capable. Adjusted for inflation, it's also an unparalleled bargain.

Gensets — especially those with the capacities and features of interest to hams — have taken a similar path. Generators that were once as loud as a fleet of garden tractors are now whisper quiet. Those heavy, bulky monstrosities are now small and lightweight enough to be carried around by your 12 year old nephew (a perfect introduction to FD!). And if you've ever fried your logging computer, your transceiver or your flat panel TV because your genset decided to output a funky 160 V, 50 Hz "modified sine wave," you'll love the rock solid ac that comes out of a modern inverter generator. Cost wise, modern generators haven't *quite* kept pace with their ham radio counterparts, but even with the tremendous performance improvements they're still about the same price — or less — than they were 10 years ago.

The Evolution of Revolution

Essentially, a generator is a motor that's operating "backward." When you apply electricity to a motor, it turns the motor's shaft (allowing it to do useful work). If you need more rotational power, add more electricity or wind a bigger motor. (I'm skipping a lot of details here.) Take the same motor and physically rotate its shaft and it generates electricity across the same terminals used to supply power when using the motor as a motor. Turn the shaft faster and the voltage and frequency increase. Turn it slower and they decrease. To some degree,

[1]Notes appear on page 75.

Figure 1 — Modern inverter generators from Yamaha, Honda, Subaru/Robin and McCulloch (Jenn Feng Industrial). See Table 1 for a partial list of specifications.

Table 1
Specifications of the Inverter Gensets Shown in Figure 1

Make and Model	Output (W) (Surge/Cont)	Run Time (h) Full / ¼ Load	Noise Range (dBA @ 21 feet)	Engine Type	Weight (Pounds)	Price (Street)	Notes
McCulloch FDD210M0	N/A / 1800	4 / N/A	60-70	105.6 cc, 3 HP	65 (shipping)	$549	a,b,c
Honda EU2000i	2000 / 1600	4 / 15	53-59	100 cc, OHC	46	$999	a,b,d
Yamaha EF2400iS	2400 / 2000	N/A / 8.6	53-58	171 cc, OHV	70	$1069	a,b,d
Subaru/Robin R1700i	N/A / 1650	N/A / 8.5	53-59	2.4 HP, OHV	46	$950	a,b,d

aHas 12 V dc output.
bHas "smart throttle" for better fuel economy.
cUnit is noticeably louder than more expensive units in its class.
dUnit has low oil alert/shutdown.

---

all motors are generators and all generators are motors. The differences are in the details and in the optimization for specific functions. Outside a middle school physics class, motors aren't very efficient generators, and vice versa, but you get the idea!

A "motor" that is optimized for generating electricity is an alternator — just like the one in your car. The most basic gensets use a lawnmower type gas engine to power an ac alternator, the voltage and frequency of which depends on rotational speed. Because the generator is directly coupled to the engine, the generator's rotational speed is determined by the speed of the engine. If the engine is running too fast or too slow, the voltage and frequency of the output will be off. If everything is running at or near the correct speed, the voltage and frequency of the output will be a close approximation of the power supplied by the ac mains — a 120 V ac sine wave with a frequency of 60 Hz. Most standard consumer gensets use two pole armatures that run at 3600 RPM to produce a 60 Hz sine wave.

Steady as She Goes, Mate

There are several electronic and mechanical methods used to "regulate" the ac output — to keep the voltage and frequency values as stable as possible as generator and engine speeds vary because of current loads or other factors. Remember, a standard generator *must* turn at a specific speed to maintain output regulation, so when more power is drawn from the generator, the engine must supply more torque to overcome the increased physical/magnetic resistance in the generator's core — the generator *can't* simply spin faster to supply the extra oomph.

Most gensets have engines that use mechanical or vacuum "governors" to keep the generator shaft turning at the correct speed. If the shaft slows down because of increasing generator demand, the governor "hits the gas" and draws energy stored in a heavy rotating flywheel, for example to bring (or keep) the shaft speed up to par. The opposite happens if the generator is spinning too fast.

Better Living through Electronics

In addition to mechanical and vacuum speed regulating systems, gensets that are a step up in sophistication additionally have electronic automatic voltage regulation (AVR) systems that use special windings in the generator core (and a microprocessor or circuit to monitor and control them) to help keep things steady near 120 V and 60 Hz. AVR systems can respond to short term load changes much more quickly than mechanical or vacuum governors alone. A decade ago AVR gensets were the cream of the crop. Today, they're mostly used in medium to large units that can't practically employ inverters to maintain the best level of output regulation. You'll find them in higher quality 5 to 15 kW "home backup systems" and in many recreational vehicles.

Isolate Source and Load

If you want the highest margin of safety when powering expensive computers, ham radios, TVs and other sensitive electronics, the current crop of portable (and even ultra portable) *inverter generators* is the only way to go. Some popular examples are shown in Figure 1. Available in outputs ranging from 1 to 5 kW with relative sizes illustrated in Figure 2, these marvels use one or more of the mechanical regulation systems mentioned previously, but their ultimate benefit comes from the use of a built-in ac-dc-ac inverter system that produces beautiful — if not perfect — 60 Hz sine waves at 120 V ac, with a 1% to 2% tolerance, even under varying load conditions. Some power companies can't do as well.

Instead of using two windings in the generator core, an inverter genset uses 24 or more windings to produce a high frequency ac waveform of up to 20 kHz. A solid state inverter module converts the high frequency ac to smooth dc, which is in turn converted to clean, tightly regulated 120 V ac power.

And that's not all. Most inverter gensets are compact, lightweight and *stunningly* quiet. When a handheld genset can pump out a kilowatt or two of squeaky clean ac power while making so little noise as to be drowned out by a normal conversation — and not the other way around — that's progress! No more hiding the generator in the gully and running a 200 foot extension cord to the gear. More detailed information on inverter gensets is available on the *QST* Binaries Web site.[2]

A Genset to Call Your Own

Before you buy any ham radio genset that *doesn't* have an inverter, remember that basic, inexpensive gensets are intended to power lights, saws, drills, ac motors, electric frying pans and stuff that can reliably be run on "cruddy power." Heck, some of that stuff could run just fine on 120 V dc! Or 168 V ac at 50 Hz. If you want to risk your sensitive hardware to the vagaries of a construction

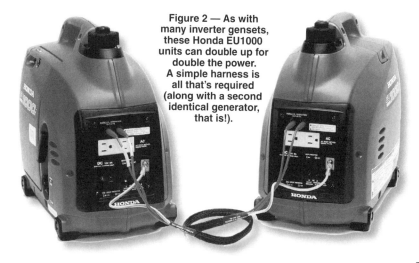

Figure 2 — As with many inverter gensets, these Honda EU1000 units can double up for double the power. A simple harness is all that's required (along with a second identical generator, that is!).

Figure 3 — It's 1, 2 and 3 — kilowatts of inverter generator power, that is. These Honda units show the relative sizes of typical 1, 2 and 3 kW inverter units industry wide.

site generator, please note the risk! Lately, at most home improvement stores you can find cute, tiny, neat looking gensets that claim to put out 750 W of ac power, weigh about as much as a bottle of laundry detergent and cost as little as $130. *Danger lurks!*

Some of these mini gensets (and many larger construction type units) have 12 V dc outputs for charging batteries. With a battery attached (and perhaps a giant capacitor) to smooth things out, you can extract 8 to 70 A of dc from 12 to 16 V. If you're tempted to use such a system, however, bring two batteries. Use one while charging and the other to protect your gear from the unknown!

While shopping at most stores that sell portable generators, don't *expect* expert help from the salespeople. Most are used to selling gensets to contractors and may not be aware of much of the information discussed in this article (although I was pleasantly surprised by the sales guy at my local Northern Tool and Supply). Do your own homework.

In addition to capacity and output regulation, other factors such as engine type, noise level, fuel options, fuel capacity, run time, size, weight, cost or connector type, may factor in your decision. Before you buy, consider additional uses for your new genset beyond Field Day. Don't forget camping, power outages, pontoon cruises and DXpeditions, for example. Although it's hard for me to imagine buying a noninverter model, you may need a solution that fits your exact requirements. Key specifications of three inverter gensets are provided in Table 1.

Capacity

Your generator must be able to safely power all of the devices that will be attached to it. That's just common sense. Simply add up the power requirements of *all* the devices, add a reasonable safety margin (25 to 30%) and choose a suitably powerful generator that meets your other requirements.

Some devices — especially electric motors — take a lot more power to start up than they do to keep running. A motor that takes 1000 W to run may take 2000 to 3000 W to start (as you'll learn when you try to power your camper's air conditioner from a tiny portable genset!). Many items don't require extra start-up power, but be sure to plan accordingly.

Always plan to have more capacity than you require or, conversely, plan to use less gear than you have capacity for. Running on the ragged edge is bad for your genset *and* your gear. Some gensets are somewhat overrated, probably for marketing purposes. (The EPA says your car will get 35 MPG, while you know it actually gets 28 MPG — this is the same kind of thing.) Give yourself a margin of safety and don't rely on built-in circuit breakers to save your gear during overloads. When operating at or beyond capacity, a genset's frequency and voltage can go wild before the current breaker pops! Some inverter gensets can be interconnected to increase capacity, as shown in Figure 3.

Size and Weight

Size and weight vary according to power output — low power units are lightweight and physically small, while beefier models are larger, weigh more and probably last longer. Watt for watt, however, most modern units are smaller and lighter than their predecessors. Models suitable for hamming typically weigh between 25 and 125 pounds.

Engines and Fuel

Low end gensets are typically powered by low tolerance, side valve engines of the type found in discount store lawnmowers. They're noisy, need frequent servicing and often die quickly. Better models have overhead valve (OHV) or overhead cam (OHC) engines, pressure lubrication, low oil shutdown, cast iron cylinder sleeves, oil filters, electronic ignition systems and even fuel injection. These features may be overkill for occasional use but desirable for more consistent power needs.

Run Time

Smaller gensets usually have smaller gas tanks, but that doesn't necessarily mean they need more frequent refueling. Some small gensets are significantly more efficient than their larger counterparts and may run for half a day while powering small loads.

As with output power, run times for many units are somewhat exaggerated and are usually spec'd for 50% loads. If you're running closer to max capacity, your run times may be seriously degraded. The opposite is also true. Typical gensets run from three to nine hours on a full tank of gas at a 50% load. If you want to power your Field Day station for a week without refueling, buy a Toyota Prius! An upcoming article will provide more information on this possibility.

Noise

Except for ham friendly inverter units — which are eerily quiet thanks to their high tech, sound dampening designs — standard gensets are almost always too loud. If Field Day is in the middle of your "back 40," the noise won't be a problem, but if you're set up in a campground or other more public space, noninverter gensets can sound like a county fair. Plan ahead to tackle the complaints.

Noise levels for many models are stated on the box, but try to test them yourself or talk to someone who owns the model you're interested in before buying. Environmental conditions, distance to the genset and the unit's physical orientation can affect perceived noise levels.

Gensets housed in special sound dampened compartments in large boats and RVs can be much quieter than typical "outside" models. They're not a free lunch, though. RV gens are expensive and heavy, use more fuel than compact models, and most don't have regulation specs comparable to inverter models.

Regulation

For hams, voltage and frequency regulation are the biggies. Any genset can safely power lightbulbs or power saws, but when it comes to computers, TVs and expensive ham radio equipment, AVR units with electronic output regulation (a minimum) and inverter gensets are highly desirable and should be used exclusively; if only for peace of mind.

Unloaded standard gensets can put out as much as 160 V ac at 64 Hz. As loads increase, frequency and voltage decrease. Under full load, output values may fall as low as 105 V at 56 Hz. Normal operating conditions are somewhere in between (as is the potential destruction of your expensive gear!).

Some hams have tried inserting uninterruptible power supplies (UPSs) between the generator and their sensitive gear. These devices are often used to maintain steady, clean ac power for computers and telecommunication equipment. As the mains voltage moves up and down, the UPS's Automatic Voltage Regulation (AVR) system bucks or boosts accordingly. The unit's internal gel cell batteries provide power to the loads if the ac mains (or your generator) go down.

In theory, this is swell. The idea is to add high quality regulation to an inexpensive generator. In practice, however, most UPSs can't handle the variation in frequency and voltage of a generator powered system. When fed by a standard generator, most UPSs constantly switch in and out of battery power mode — or don't *ever* switch back to ac power. When the UPS battery goes flat, the unit shuts off. Then, it's back to powering your sensitive gear with an insensitive generator! Not *every* UPS and *every* generator lock horns like this, but it's now much easier to get an inverter genset and be done with it. If you're determined to try it, test and retest at home before heading out to Field Day!

DC Output

Some gensets have 12 V dc outputs for charging batteries. These range from 2 A trickle chargers to 100 A powerhouses (good for starting tractors or MacGyver style improvised welding). Typical outputs run about 10 to 15 A. As with the ac outputs, be sure to test the dc outputs for voltage stability (under load if possible) and ripple. Car batteries aren't too fussy about a little ripple in the charging circuit, but your radio might not like it at all!

Miscellaneous

Other considerations include outlets (120 V, 240 V and dc output), circuit breakers (standard or ground fault interrupter type), fuel level gauges, handles (one or two), favorite brands, warranties, starters (pull or electric), wheels, handles or whatever you require.

Setup, Safety and Testing

Before starting the engine, read the user manual — at least twice — cover to cover. Carefully follow the instructions regarding engine oil, throttle and choke settings (if any). Be sure you understand how the unit operates and how to use the receptacles, circuit breakers and connectors.

Make sure the area is clean, dry and unobstructed. Gensets should *always* be set up *outdoors*. Do not operate gas powered engines in closed spaces, inside passenger vans, inside covered pickup beds, etc. If rain is a possibility, set up an appropriate canopy or other *outdoor protective structure*. Operating generators and electrical devices in the rain or snow can be dangerous. Keep the generator and any attached cords dry!

Exhaust systems can get hot enough to ignite certain materials. Keep the unit several feet away from buildings, and keep the gas can (and other flammable stuff) at a safe distance. Don't touch hot engines or mufflers!

When refueling, shut down the generator and let things cool off for a few minutes. Don't smoke, and don't spill gasoline onto hot engine parts. A flash fire or explosion may result. Keep a small fire extinguisher nearby. If you refuel at night, use a light source that isn't powered by the generator and can't ignite the gasoline.

Testing

Before starting (or restarting) the engine, *disconnect* all electrical loads. Starting the unit while loads are connected may not damage the generator, but your solid state devices may not be so lucky.

After the engine has warmed up and stabilized, test the output voltage (and frequency), if possible, *before* connecting loads.

Because unloaded values may differ from loaded values, be sure to test your generator under load (using high wattage quartz lights or an electric heater as appropriate). Notice that when you turn on a hefty load, your generator will "hunt" a bit as the engine stabilizes. Measure ac voltage and frequency again to see what the power conditions will be like under load. See your unit's user manual or contact the manufacturer if adjustments are required.

Safety Grounds and Field Operation

Before we can connect *real* electrical loads in a Field Day situation we need to choose a grounding method — a real controversy among campers, RVers and home power enthusiasts.

To complicate matters, most gensets have ac generator grounds that are connected to their metal frames, but some units do not bond the ac neutral wires to the ac ground wires (as in typical house wiring). Although they will probably safely power your ham station all day long, units with unbonded neutrals may appear defective if tested with a standard ac outlet polarity tester.

Some users religiously drive copper ground rods into the ground or connect the metal frames of their generators to suitable existing grounds, while others vigorously oppose this method and let their gens float with respect to earth ground (arguing that if the generator isn't connected to the earth, you can't complete the path to earth ground with your body should you encounter a bare wire powered by the generator; no path, no shock). Some user manuals insist on the ground connection, while others don't.

You can follow your unit's user manual, check your local electric code, choose a grounding method based on personal preference or expert advice, or do further research. Either method may offer better protection depending on exact circumstances.

Regardless of the grounding method you choose, a few electrical safety rules remain the same. Your extension cords *must* have intact, waterproof insulation, three "prongs" and three wires, and must be sized according to loads and cable runs. Use 14 to 16 gauge, three wire extension cords for low wattage runs of 100 feet or less. For high wattage loads, use heavier 12 gauge, three wire cords designed for air compressors, air conditioners or RV service feeds. If you use long extension cords to power heavy loads, you may damage your generator or your radio gear. When it comes to power cords, think *big*. Try to position extension cords so they won't be tripped over or run over by vehicles. And don't run electrical cords through standing water or over wet, sloppy terrain.

During Field Day operations, try to let all operators know when the generator will be shut down for refueling so radio and computer gear can be shut down in a civilized manner. Keep the loads disconnected at the generator until the generator has been refueled and restarted.

Conclusion

Modern inverter gensets are the perfect companions for today's computer powered Amateur Radio gear. Getting on the air when the mains go down — or when there are no ac mains — has never been easier, safer and quieter. Until even smaller, completely silent "field supplies" powered by mini fuel cells become affordable, that is!

Notes
[1]K. Kleinschmidt, NT0Z, "How to Choose and Use a Portable Power Generator," *QST*, Jun 1999, pp 59-61.
[2]**www.arrl.org/files/qst-binaries/**.

The author, a QST *editor from 1988 through 1994, now lives in a Southeastern Minnesota condo and, although he wrote* Stealth Amateur Radio, *he never thought he'd actually have to operate on the sly. Now that he's written this article, he's hoping for an inverter generator to broaden his field activities, ham radio and otherwise. You can reach him at 6126 S Pointe Dr SW, Rochester, MN 55902 or at* **kirk@cloudnet.com**.

Public Relations

- Press Release
- Publicity Tip Sheet

LOCAL RADIO ENTHUSIASTS SHARE THEIR PASSION FOR TECHNOLOGY AND PUBLIC SERVICE DURING AMATEUR RADIO WEEK

(PUT THIS RELEASE ON YOUR CLUB'S LETTERHEAD)

For More Information Contact:
(Name of your local contact)
(Contact's phone number)

(*TOWN, State, date*) -- On-air demonstrations, satellite communications and emergency preparedness activities are on tap for members of the (your club name) as they gear up to celebrate Amateur Radio Week, June 21-26, 2010.

During Amateur Radio Week, enthusiasts will put on demonstrations, give talks to community groups and take part in other activities to raise awareness about Amateur Radio. The week culminates with the annual preparedness exercise called "Field Day," June 26 and 27. Field Day is sponsored by the American Radio Relay League (ARRL), the national association for Amateur Radio.

During Field Day, operators set up in local parks, at shopping malls, or even in their own backyards, and get on the air using generators or battery power. Field Day was designed to test operators' abilities to set up and operate portable stations under emergency conditions such as the loss of electricity. "We want the community to know that in the event of an emergency, we will be ready to assist in any way we can, says (your club spokesperson). "While people often think that cell phones or other communications technologies have replaced ham radio, we can still provide an important communications service that others can't."

Field Day is a serious test of skill, but it is also a contest for fun and the largest "on-air" operating event each year. During the weekend, radio operators try to contact as many other Field Day stations as possible. More than 30,000 Amateur Radio operators across the country and Canada participated in last year's event.

The (your club name) will hold Field Day at (site location) and will be operating from (date/time to date/time). "We hope that anyone who is interested in seeing what Amateur Radio is all about will come out for Field Day," says (your club spokesperson).

Today there are more than 670,000 Amateur Radio operators in the United States and more than 2.5 million worldwide. To find out more about Amateur Radio or how you can get started, contact (your club contact) at (telephone number and e-mail address). Information is also available from the American Radio Relay League, 225 Main Street, Newington, CT 06111 or by calling 1-800-32-NEW HAM. The URL for the ARRL's home page on the World Wide Web is www.arrl.org.

###

Amateur Radio Week Publicity Tip Sheet

Ideas to help you promote Amateur Radio Week and Field Day

1. Retype the enclosed news release onto your club letterhead. Be sure to fill in your club's name and contact in the appropriate blanks. Feel free to lift any text from this release and add it to one you've already created.

2. Mail or fax the news release to the city editor of your local paper, radio and television stations two to three weeks in advance of the time you want it to run. You may wish to follow-up with a telephone call within a few days to see that it was received and offer any additional information.

3. Approach your local cable TV and radio stations with ARRL public service announcements. If they're aired, you'll get good exposure for Amateur Radio and your club prior to Amateur Radio Week and Field Day activities.

4. Give the enclosed backgrounders to reporters looking for more information or use them to help you write up your publicity materials.

5. If a local editor or reporter expresses interest in ham radio or local classes, consider inviting him or her to take a course and write about it.

6. Consider developing informational handouts that can be left in local radio/electronics stores.

7. Consider posting attractive, easy-to-read notices in prominent places: libraries, supermarkets, radio/electronics stores, schools, etc.

8. Put a local reporter on your club newsletter mailing list.

9. Set up a station in a public place. Shopping malls are great. Have plenty of handouts on hand, and be sure a sign identifies who you are and what you're doing.

10. Call local broadcast radio talk shows and volunteer to be a guest. Be sure to propose a thought-provoking topic.

11. Write a letter to the editor of your local paper and invite readers to visit your Field Day site and learn more about Amateur Radio.

12. Hands-on involvement builds commitment. So consider ways to let others "try out" the equipment. Third party operation with other members of your club would be a thrill for local school children, for example.

13. Volunteer to speak on the subject of Amateur Radio at a local Rotary, or other service club meeting.

14. If your club has a Web page, make sure you pass the URL on to the media you are working with. If not, or for more information on ham radio give ours (http://www.arrl.org/).

Steve Ford, WB8IMY

Amateur Satellites

Hams were present at the dawn of the Space Age, creating the first amateur satellite in 1961, and we've been active on the "final frontier" ever since. Even so, satellite-active hams compose a relatively small segment of our hobby, primarily because of an unfortunate fiction that has been circulating for many years — the myth that operating through amateur satellites is difficult and expensive.

Like any other facet of Amateur Radio, satellite hamming is as expensive as you allow it to become. If you want to equip your home with a satellite communication station that would make a NASA engineer blush, it will be expensive. If you want to simply communicate with a few low-Earth-orbiting birds using less-than-state-of-the-art gear, a satellite station is no more expensive than a typical HF or VHF setup.

What about difficulty? Prior to 1982, hams calculated satellite orbits using an arcane manual method that many people found unfathomable. In truth, the manual method taught you a great deal about orbital mechanics, but it was viewed by some as being too difficult. Today computers do all of the calculations for you and display the results in easy to understand formats (more about this later). Satellite equipment also has become much easier for the average ham to use.

SATELLITES: ORBITING RELAY STATIONS

Most amateurs are familiar with repeater stations that retransmit signals to provide coverage over wide areas. Repeaters achieve this by listening for signals on one frequency and immediately retransmitting whatever they hear on another frequency. Thanks to repeaters, small, low-power radios can communicate over thousands of square miles.

This is essentially the function of an amateur satellite as well. Of course, while a repeater antenna may be as much as a few thousand meters above the surrounding terrain, the satellite is hundreds or thousands of kilometers above the surface of the Earth. The area of the Earth that the satellite's signals can reach is therefore much larger than the coverage area of even the best Earth-bound repeaters. It is this characteristic of satellites that makes them attractive for communication. Most amateur satellites act either as analog repeaters, retransmitting CW and voice signals exactly as they are received, or as packet store-and-forward systems that receive whole messages from ground stations for later relay.

Linear Transponders

Several analog satellites are equipped with *linear transponders*. These are devices that retransmit across a band of frequencies, usually 50 to 100 kHz wide, known as the *passband*. Since the linear transponder retransmits the entire band, many signals may be retransmitted simultaneously. For example, if four SSB signals (each separated by 20 kHz) were transmitted to the satellite, the satellite would retransmit all four signals — still separated by 20 kHz each. Just like a terrestrial repeater, the retransmission takes place on a frequency that is different from the one on which the signals originally were received.

In the case of amateur satellites, the difference between the transmit and receive frequencies is similar to what you might encounter on a crossband terrestrial repeater. In other words, retransmission occurs on a different *band* from the original signal. For example, a transmission received by the satellite on 2 meters might be retransmitted on 70 cm. This crossband operation allows the use of simple filters in the satellite to keep its transmitter from interfering with its receiver.

Some linear transponders invert the uplink signal. In other words, if you transmit to the satellite at the *bottom* of the uplink passband, your signal will appear at the *top* of the downlink passband. In addition, if you transmit in upper sideband (USB), your downlink signal will be in lower sideband (LSB). Transceivers designed for amateur satellites usually include features that cope with this confusing flip-flop.

Linear transponders can repeat any type of signal, but those used by amateur satellites are primarily

Two Field Day operators rise early to get their chance at working a satellite.

79

> ### Don't Be a Transponder Hog
>
> A linear transponder retransmits a faithful reproduction of the signals received in the uplink passband. This means that the loudest signal received will be the loudest signal retransmitted. If a received signal is so strong that retransmitting it would cause an overload, automatic gain control (AGC) within the transponder reduces the transponder's amplification. However, this reduces all of the signals passing through it!
>
> Strong signals don't receive any benefit from being loud since the downlink is "maxed out" anyway. Instead, all of the other users are disrupted because their signals are reduced. Avoiding this condition is just common sense — and good operating practice. It's easy, too: just note the signal level of the satellite's beacon transmission and adjust your transmit power to a level just sufficient to make your downlink appear at the same level as the beacon. Any additional power is too much!

designed for SSB and CW. The reason has to do with the problem of generating power in space. Amateur satellites rely on batteries that are recharged by solar cells. "Space rated" solar arrays and batteries are expensive. They are also heavy and tend to take up a substantial amount of space. Thanks to meager funding, hams don't have the luxury of launching satellites with large power systems such as those used by commercial birds. We have to do the best we can within a much more limited "power budget."

So what does this have to do with SSB or any other mode?

Think of *duty cycle* — the amount of time that a transmitter operates at full output. With SSB and CW the duty cycle is quite low. A linear transponder can retransmit many SSB and CW signals while still operating within the power generating limitations of an amateur satellite. It hardly breaks a sweat.

Now consider FM. An FM transmitter operates at a 100% duty cycle, which means it is generating its *full output* continuously during every transmission. Imagine how much power a linear transponder would need to retransmit, say, a dozen FM signals — all demanding 100% output!

Having said all that, there *are* a few FM repeater satellites, and we'll discuss them in this chapter. However, these are very low-power satellites (typically less than 1 W output) and they do not use linear transponders. They retransmit only one signal at a time, not many signals simultaneously.

FINDING A SATELLITE

Before you can communicate through a satellite, you have to know when it is available. Computers make this task very easy. First, however, we need to understand a little bit about how ham satellites behave in orbit.

Amateur satellites, unlike many commercial and military spacecraft, do not travel in geostationary orbits. Satellites in geostationary orbits cruise above the Earth's equator at an altitude of about 22,000 miles. From this vantage point the satellites can "see" almost half of our planet. Their speed in orbit matches the rotational speed of the Earth itself, so the satellites appear to be "parked" at fixed positions in the sky.

They are available to send and receive signals 24 hours a day over an enormous area.

Of course, amateur satellites *could* be placed in geostationary orbits. The problem isn't one of physics; it's money and politics. Placing a satellite in geostationary orbit and keeping it there costs a great deal of money — more than any one amateur satellite organization can afford. An amateur satellite group could ask similar groups in other areas of the world to contribute funds to a geostationary satellite project, but why should they? Would you contribute large sums of money to a satellite that may never "see" your part of the world? Unless you are blessed with phenomenal generosity, it would seem unlikely!

Instead, all amateur satellites are either low-Earth orbiters (LEOs), or they travel in high, elongated orbits. Either way, they are not in fixed positions in the sky. Their positions relative to your station change constantly as the satellites zip around the Earth. This means that you need to predict when satellites will appear in your area, as well as what paths they'll take as they move across your local sky.

A bare-bones satellite-tracking program will provide a schedule for any satellite you choose. A very simple schedule might look something like this:

Date	Time	Azimuth	Elevation
10 OCT 07	1200	149°	4°
10 OCT 07	1201	147°	8°
10 OCT 07	1202	144°	13°
10 OCT 07	1203	139°	20°

The date column is obvious: 10 October 2007. The time is usually expressed in UTC. This particular satellite will appear above your horizon beginning at 1200 UTC. The bird will "rise" at an azimuth of 149°, or approximately southeast of your station. The elevation refers to the satellite's position above your horizon in degrees — the higher the better. A zero-degree elevation is right on the horizon; 90° is directly overhead.

By looking at this schedule you can see that the satellite will appear in your southeastern sky at 1200 UTC and will rise quickly to an elevation of 20° by 1203. The satellite's path will curve further to the east as it rises. Notice how the azimuth shifts from 149° at 1200 UTC to 139° at 1203.

The more sophisticated the software, the more information it usually provides in the schedule table. The software may also display the satellite's position graphically as a moving object superimposed on a map of the world. Some of the displays used by satellite prediction software are visually stunning! See **Fig 10-1** for an example.

Table 10-1
Consistently Active Amateur Satellites: Frequencies and Modes

Satellite	Uplink (MHz)	Downlink (MHz)
SSB/CW		
VuSAT-OSCAR 52	435.220-435.280	145.870-145.930
HOPE-OSCAR 68	145.925-145.975	435.765-435.715
FM Voice Repeaters		
AMSAT-OSCAR 51	145.920	435.300
OSCAR 67	145.875	435.345 233.6 Hz CTCSS

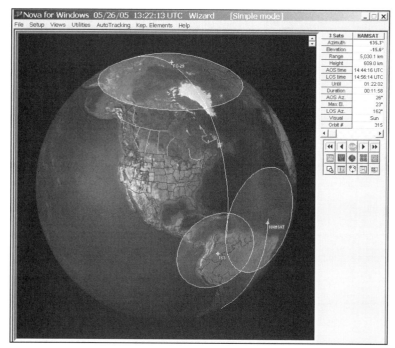

Fig 10-1 — This is one of several tracking displays provided by the *Nova* satellite-tracking software (available from AMSAT).

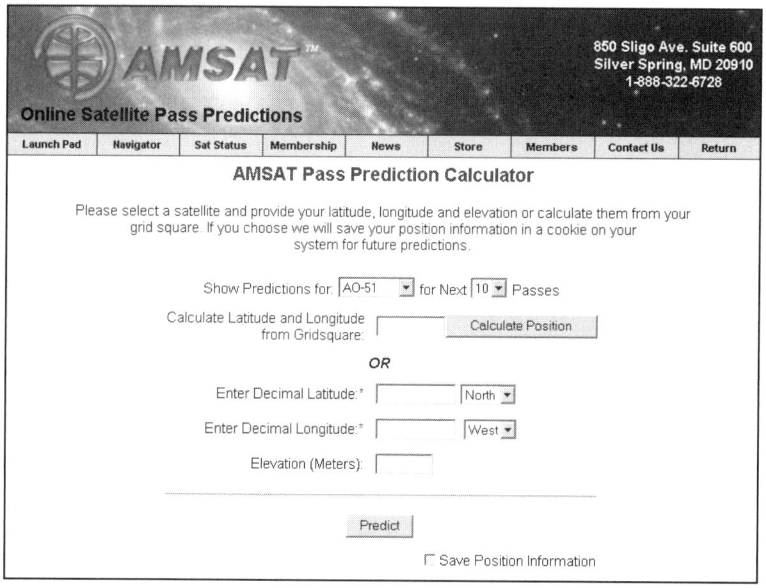

Fig 10-2 — You can also find satellite pass predictions on the Web at www.amsat.org/amsat-new/tools/predict/.

Satellite prediction software is widely available on the Web. Some of the simpler programs are freeware. A great place to start is the AMSAT-NA site at **www.amsat.org**. They have the largest collection of satellite software for just about any computer you can imagine. AMSAT software isn't free, but the cost is reasonable and the funds support amateur satellite programs.

You can even dispense with software altogether and use the pass predictor on the AMSAT-NA Web site to set your schedule. You'll find it at **www.amsat.org/amsat-new/tools/predict/** (see **Fig 10-2**).

Whichever approach you choose, you must provide two key pieces of information before you can obtain pass predictions:

1) *Your position.* The software must have your latitude and longitude before it can crank out predictions for your station. The good news is that your position information doesn't need to be extremely accurate. Just find out the latitude and longitude and plug it into the program. You can obtain the longitude and latitude of your town by calling your public library or nearest airport. It is also available on the Web at **geonames.usgs.gov/pls/gnis/web_query.gnis_web_query_form**.

2) *Orbital elements.* This is the information that describes the orbits of the satellites. You can find orbital elements (often referred to as *Keplerian elements*) at the AMSAT Web site, and through many other sources on the Internet. You need to update the elements every few months. Many satellite programs will automatically read in the elements if they are provided as ASCII text files. The less sophisticated programs will require you to enter them by hand. The automatic-update software helps to avoid typing mistakes with manual entries. Note that if you use the AMSAT-NA online satellite pass predictor, you do *not* need to supply orbital elements.

Table 10-1 shows the amateur satellites that were consistently active when this book went to press. Check **www.amsat.org** for the latest status of these and other birds.

GETTING STARTED WITH FM REPEATER SATELLITES

Would you like to operate through an FM repeater with a coverage area that spans an entire continent? Then check out the AMSAT-OSCAR 51 FM repeater satellite (**Fig 10-3**). From its near-polar orbit at altitudes of approximately 500 miles, this satellite can hear stations within a radius of 2000 miles in all directions. OSCAR 51's FM repeater has one channel with an uplink frequency of 145.920 MHz and a downlink frequency of 435.300 MHz, both plus/minus the Doppler effect (sidebar "Down with the Doppler").

You can operate through OSCAR 51 with a basic dual-band FM transceiver with 25 W output or more. Assuming that the transceiver is reasonably sensitive, you can use an omnidirectional antenna such as a dual-band collinear ground plane or something similar. Some amateurs even have managed to work it with handheld transceivers, but to reach an FM satellite with a handheld you'll need to use a multi-element directional antenna. Of course, this means that you'll have to aim your antenna at the satellites as they cross overhead.

Start by booting up your satellite tracking software, or grabbing a schedule from the AMSAT-NA online predictor. Check for a pass where the satellite rises at least 45° above your horizon. As with all satellites, the higher the elevation, the better. If you plan to operate outdoors or away from home, either print the schedule or jot down the times on a piece of

Fig 10-3 — This solar-cell-covered box is OSCAR 51, a popular FM repeater satellite.

Fig 10-4 — A view of OSCAR 52 just prior to launch. (It is the small, dark cube below and to the right of the large commercial satellite above.)

Down with the Doppler

The relative motion between you and the satellite causes *Doppler shifting* of signals. As the satellite moves toward you, the frequency of the downlink signals will increase by a small amount. As the satellite passes overhead and starts to move away from you, there will be a rapid drop in frequency of a few kilohertz, much the same way as the tone of a car horn or a train whistle drops as the vehicle moves past the observer.

The Doppler effect is different for stations located at different distances from the satellite because the relative velocity of the satellite with respect to the observer is dependent on the observer's distance from the satellite. The result is that signals passing through the satellite transponder shift slowly around the calculated downlink frequency. Your job is to tune your uplink transmitter — *not your receiver* — to compensate for Doppler shifting and keep your frequency relatively stable on the downlink. That's why it is helpful to hear your own signal coming through the satellite. If you and the station you're talking to both compensate correctly, your conversation will stay at one frequency on the downlink throughout the pass. If you don't compensate, you will drift through the downlink passband as you attempt to "follow" each other's signals. This is highly annoying to others using the satellite because your drifting signals may drift into their conversations.

Doppler shift through a transponder becomes the sum of the Doppler shifts of both the uplink and downlink signals. In the case of an inverting type transponder (as in OSCAR 52), a Doppler-shifted increase in the uplink frequency causes a corresponding decrease in downlink frequency, so the resultant Doppler shift is the *difference* of the Doppler shifts, rather than the *sum*. The shifts tend to cancel.

---

scrap paper that you carry with you.

When the satellite comes into range, you'll be receiving its signal about 10 kHz higher than the designated downlink frequency, thanks to Doppler shifting. So, begin listening there. At about the midpoint of the pass you'll need to shift your receiver down 10 kHz or more. As the satellite is heading away, you may wind up stepping down to as much as 10 kHz below the downlink frequency. Some operators program these frequencies into memory channels so that they can compensate for Doppler shift at the push of a button.

These satellites behave just like terrestrial FM repeaters. Only one person at a time can talk. If two or more people transmit simultaneously, the result is garbled audio or a squealing sound on the output. The trick is to take turns and keep the conversation short. Even the best passes will only give you about 15 minutes to use the satellite. If you strike up a conversation, don't forget that there are others waiting to use the bird.

It is also a good idea to check the OSCAR 51 schedule on the Web at www.amsat.org/amsat-new/echo/ControlTeam.php. The bird occasionally changes its operating mode for experiments.

The FM repeater satellites are a great way to get started. They are easy to hear and easy to use. Once you get your feet wet, however, you'll probably wish you could access a satellite that wasn't so crowded, where you could chat for as long as the bird is in range.

MOVING UP TO VUSAT-OSCAR 52 AND HOPE-OSCAR 68

The FM repeater satellites are a fun way to get your feet wet, but soon you'll want to graduate to satellites that offer the luxury of longer conversations and minimal interference — such as VuSAT-OSCAR 52 or HOPE-OSCAR 68. Shown on **Fig 10-4**, OSCAR 52 is a linear-transponder bird that listens on 70 cm and repeats everything it hears on the 2-meter band. (This is called "Mode U/V" in satellite jargon. See the sidebar, "Satellite Modes Demystified.")

OSCAR 52/68 — Hardware

To set up a station for OSCAR 52 or 68 you'll need either (A) an SSB/CW transceiver that can operate on 2 meters and 70 cm in full duplex, (B) an SSB/CW transmitter on 70 cm and a separate 2-meter receiver or (C) an SSB/CW transmitter on

The modest satellite antennas of VE3VRW are mounted on a small az/el rotator.

70 cm and a 2-meter receive converter attached to a 10 meter receiver. Let's look at each option.

(A) VHF/UHF all-mode transceivers designed for satellite operating are available, although the newer radios usually come with price tags around $1500. Despite the cost, these radios are good investments if you intend to expand your satellite operating.

(B) Many popular HF/VHF transceivers today feature 2-meter SSB/CW capability, so using one of those radios for your downlink isn't a problem. SSB/CW transmitters or transceivers for the 70-cm uplink are less common, but some HF/VHF+ transceivers include this band.

(C) The least expensive way to enjoy OSCAR 52 or 68 is to use a UHF SSB/CW transceiver for the uplink and a 2-meter receive converter with a separate HF (10-meter) receiver for the downlink. These receive converters are relatively inexpensive and they are available from a number of vendors.

You can use omnidirectional antennas such as groundplanes, eggbeaters and others to work OSCAR 52 or 68, but the results may be disappointing. If you choose this approach, you may find that you need to run substantial power on the 70-cm uplink (about 100 W) and a preamplifier on the 2-meter downlink antenna to achieve consistent results.

If you can afford it, the better choices by far are directional antennas on 2 meters and 70 cm (such as Yagis). With directional antennas you'll enjoy solid signals throughout each pass. Of course, this raises the issue of how you will track the satellites with your antennas — either by hand or by using an az/el (azimuth and elevation) rotator (**Fig 10-5**). The decision to invest in an az/el rotator will probably hinge on whether you plan to try other VHF and/or UHF satellites in the future.

If you are using directional antennas, you will find that you'll need to readjust their positions every couple of minutes. Attempting to do this manually while carrying on a conversation (and compensating for Doppler) can be quite a juggling act. In this situation, an az/el rotator is particularly attractive.

And if you're using satellite-tracking software that supports rotator control, a computer rotator control interface is also worth considering. This turns the job of antenna aiming over to your computer, leaving you free to concentrate on the radios.

OSCAR 52/68 — Operating

When you're ready to make contacts through OSCAR 52 or 68, you'll discover an interesting twist: this satellite uses an *inverting transponder*. This means that the downlink passbands are inverted mirror images of the uplink passbands. If you transmit to the satellite in the *lower* portions of the uplink passband, your signal will appear on the *upper* portion of the downlink passband! In addition, if you transmit to the satellites in *lower* sideband (LSB), your signal will be repeated on the downlink in *upper* sideband (USB).

Here's an example: You're transmitting at: 435.230 MHz LSB and you hear your signal from the satellite at: 145.920 MHz USB. Notice the "upside down" relationship? Your LSB signal on the uplink has become USB on the downlink. Your downlink signal appears 10 kHz *below* the *top* end of the downlink passband because you transmitted 10 kHz *above* the *bottom* end of the uplink passband.

Many VHF/UHF transceivers that are intended for satellite operating include a feature that allows you to lock the uplink and downlink VFOs so that they'll track in reverse. Using our example above, as you begin tuning the downlink VFO *downward* from 145.930 to 145.870 MHz, the uplink VFO will automatically increment *upward* from 435.220 to 435.280 MHz. This takes a lot of the confusion out of working satellites that use inverting transponders. The convention with OSCAR 52 and 68 is to transmit in LSB so that your signal is USB on the downlink.

Wear a pair of headphones so that you can monitor your own signal on the downlink. Find a quiet spot in the passband

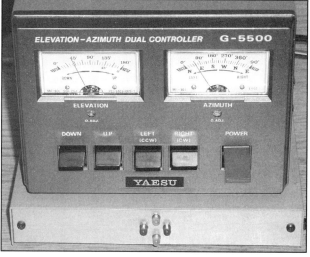

Fig 10-5 — An azimuth/elevation (az/el) rotator has two motors. One moves the antenna array from side to side, while the other moves it up and down. The controller usually has two sets of controls and two position displays so you know exactly where the antenna is pointed. Some controllers have computer interfaces so that tracking software can automatically update antenna position during a pass.

and say (or send) your call sign repeatedly. Tune your downlink receiver above and below where you think your signal should be. In time you'll learn the knack and be able to "find yourself" quickly. If you're using a transceiver with reverse-tracking VFOs, it is best to disable the auto-tracking function until you locate your signal.

If you hear someone calling CQ, do a rough calculation of the necessary uplink frequency and begin calling the other station repeatedly, tuning your *uplink* radio until you hear yourself (unlock your auto-tracking VFOs if you have them). The other station will no doubt be tuning around as well. Once the conversation is established, change only your *downlink* frequency setting to compensate for the Doppler Effect.

Considering the substantial Doppler frequency shifting that takes place on the downlinks, it's best to use short exchanges to minimize the amount of uplink re-tuning necessary at the beginning of each transmission.

Keep your transmissions short. That'll make it easier for the other station to compensate for Doppler. Operating on OSCAR 52 and 68

Steve Ford, WB8IMY

HF Digital Communications

We tend to think of the HF bands in terms of voice (SSB and AM) and CW activity. With the ubiquity of the personal computer, however, digital communication has seen an increasing HF presence as well. This has been particularly true for amateurs who are forced to operate in restricted situations using such things as hidden antennas and low output power. They've discovered that a few watts with a digital mode can still take them considerable distances, even around the world.

EVOLUTION AND REVOLUTION

Radioteletype, better known as *RTTY*, was one of the first Amateur Radio digital modes. Hams began using it in the late 1940s and the technology remained essentially unchanged for more than 30 years. If you had visited an amateur RTTY station prior to about 1980, you probably would have seen a hulking mechanical teletype machine, complete with rolls of yellow paper. The teletype would be connected to the transceiver through an interface known as a *TU*, or *terminal unit*. An oscilloscope would probably have graced the layout as well. Oscilloscopes were necessary for proper tuning of the received signal. The entire RTTY conversation was printed on, and had to be read from, an endless river of paper.

The first serious change occurred when affordable microprocessor technology appeared in the late 1970s. That's when we started to see TUs that included their own self-contained keyboards and displays, making the mechanical teletype (and all that paper!) obsolete. When personal computers debuted in the early 1980s, they became perfect companions for the new TUs. The PC functioned as a "dumb terminal," displaying the received data *from* the TU and sending data *to* the TU for transmission. Some TUs of this era offered ultra-sharp receive filters that allowed hams to copy weak signals even in the midst of horrendous interference. TU models such as the HAL Communications ST-8000 were renowned for their performance. Some RTTY operators still use ST-8000s to this day.

In the late 1980s, conventional terminal units began to yield to sophisticated devices known as *multimode controllers*. As the name suggests, these compact units handle several different digital modes in one package, typically RTTY, packet, AMTOR and PACTOR. The Kantronics KAM, AEA PK-232 and MFJ-1278 are well-known examples. Like TUs, multimode controllers are standalone devices that communicate with your personal computer. When using a multimode controller, your computer is, once again, acting as a dumb terminal — all of the heavy lifting is being done by the controller and its self-contained software known as *firmware*. These controllers functioned as radio modulators/demodulators, or *modems*, converting data to modulated audio signals for transmission and also converting received signals back to data for display.

The winds of change began to stir again in the early 1990s when sound cards appeared as accessories for personal computers. At first, sound cards were used for little more than…well… *sound*. They made it possible for computer users to enjoy music, sound effects in their computer games and other applications. But as sound cards became more powerful, hams began to realize their potential. They discovered that with the right software a sound card could function as a modem, too.

Peter Martinez, G3PLX, exploited the potential of the sound-card-as-modem when he created an entirely new amateur digital mode known as *PSK31*. It was not only a new mode, but a new way of using sound cards. The sound card not only decoded the received signal, it created the transmitted signal too. The only piece of hardware necessary (other than the transceiver, of course!) was a simple interface that allowed the computer to switch the radio from receive to transmit, and vice versa.

In the years that followed, sound cards became more powerful and versatile. They also began showing up as sound *chips* on personal computer motherboards and inside deluxe interfaces. Hams responded by devel-

Jose, YV6BTF, is often heard on RTTY.

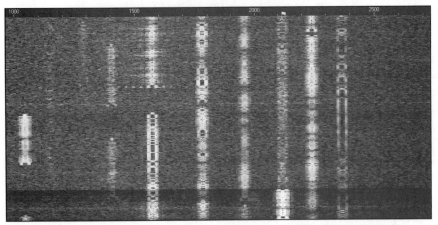

Fig 8-1 — A typical waterfall display. This one shows several PSK31 signals (represented by the vertical lines).

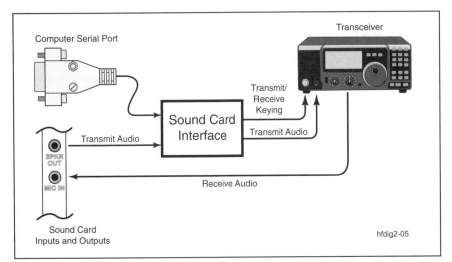

Fig 8-2 — The sound card interface installs between your computer and your HF transceiver. Its primary function is to allow your computer to switch your radio between transmit and receive, but many models perform additional functions such as providing audio isolation.

oping more new digital modes to go with them. At the time this book went to press, there were well over a dozen sound-card HF digital modes with more in their pre-release stages.

The hardware multimode controllers are still with us, but they are primarily used for modes like PACTOR that require more processing muscle and precise timing than a typical personal computer can provide on its own. All of the other Amateur Radio HF digital modes have gone the way of the sound card.

YOUR HF DIGITAL STATION

You can explore most HF digital modes with little more than a sound-card-equipped computer and an SSB transceiver.

The computer doesn't need to be particularly powerful unless you intend to explore processor-intensive applications such as digital voice or software-defined radios. A 1 GHz processor and 500 Mbytes of RAM should be adequate for most purposes. Low-priced computers available today are far more powerful than this and you can pick up used computers on eBay and elsewhere that meet this requirement at a cost of less than $500.

Your software selections are easy and affordable, too. There are excellent multimode applications that, as the term implies, provide many HF digital modes in a single program. Among the most popular are...

MixW for *Windows* at **www.mixw.net**

MultiPSK for *Windows* at **f6cte.free.fr/index_anglais.htm**

Multimode for *MacOS* at **www.black-catsystems.com**

Cocoamodem for *MacOS* at **homepage.mac.com/chen/index.html**

Fldigi for *Linux* at **www.w1hkj.com/Fldigi.html**

All of these sound card applications, and many others, work by taking the receive audio from your radio and creating a visual representation of the detected signals on your computer monitor, usually in what is known as a *waterfall*. You can see an example of a typical waterfall display in **Fig 8-1**. To decode a signal, all you have to do is click your mouse cursor on the signal's pattern in the waterfall.

Finally, you'll need a sound card interface to tie everything together. The sound card interface acts as the link between your computer and your transceiver (see **Fig 8-2**). It's most basic function is to allow your computer to switch your radio between transmit and receive. Many interfaces also allow you to set the transmit and receive audio levels. Interface manufacturers include MFJ (**www.mfjenterprises.com**), West Mountain Radio (**www.westmountainradio.com**), microHAM (**www.microham.com**), TigerTronics (**www.tigertronics.com**), MixW RigExpert (**www.rigexpert.net**) and more. **Fig 8-3** shows a typical unit. See the advertising pages of *QST* magazine for more information on specific devices.

Fig 8-3 — Some transceiver interfaces such as the MixW RigExpert Standard include extra features. This model has a built-in sound card chip set for digital modes and even includes a CW keyer.

Sound card interfaces connect to your computer through the serial (COM) port or via a USB port. You can make the radio connections at the microphone and external speaker jack, or more conveniently, at the rear-panel "accessory" jack. The interface manual will explain all this in detail.

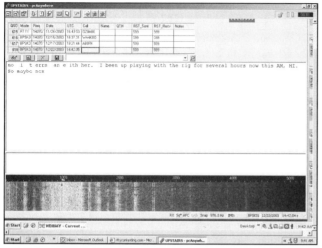

MixW multimode HF digital software.

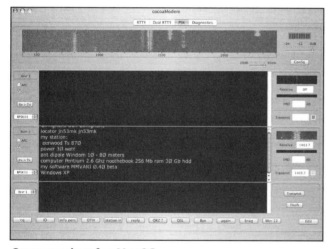

Cocoamodem for *MacOS*.

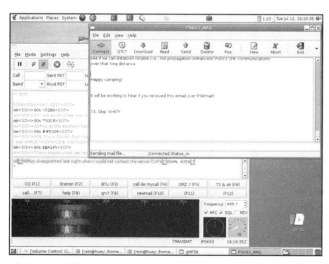

Fldigi for *Linux*.

Which Sound Card is Best?

This is one of the most popular questions among HF digital operators. After all, the sound device is second only to the radio as the most critical link in the performance chain. A poor sound device will bury weak signals in noise of its own making and will potentially distort your transmit audio as well.

Before you dash out to purchase a costly high-end sound card, ask yourself an important question: How do you intend to operate? If you have a modest station and intend to enjoy casual chats and a bit of DXing, save your money. An inexpensive sound card, or the sound chipset that is probably on your computer's motherboard, is adequate for the task. There is little point in investing in a luxury sound card if you lack the radio or antennas to hear weak signals to begin with, or if they cannot hear you.

On the other hand, if you own the station hardware necessary to be competitive in digital DX hunting or contesting, a

Keeping Everything on the Level

One of the perennial problems with sound devices is the fact that optimum audio levels have a tendency to differ depending on the software you are using. The sound card input and output levels you set correctly for one application may be wildly wrong for another. And what happens when another family member stops by the computer to listen to music or play a game? They're likely to change the audio levels to whatever suits them. When it comes time to use the computer again for HF digital operating, you may get an unpleasant surprise.

QuickMix for *Windows*.

The common sense answer is to simply check and reset the audio levels before you operate. However, what if you can't quite remember the settings? At the very least, you'll have to go back through and tweak for best receive and transmit audio. Fortunately, there are some software applications that will "remember" your audio settings and reapply them for you. One example is *QuickMix* for *Windows*, which is available free on the Web at www.ptpart.co.uk/quickmix/. *QuickMix* takes a "snapshot" of your sound card settings and saves them to a file that you name according to the application (for example, "PSK31"). To use *QuickMix* you simply open the program, click the LOAD button and select the settings you need. *QuickMix* instantly adjusts your sound card according to the configuration snapshot you saved.

An even more elegant solution is the free *Sound Card Manager* for *Windows* by Roger Macdonald, W8RJ, which you can download at www.romac-software.com. Whenever you start an application (such as a piece of HF digital software), *Sound Card Manager* automatically reconfigures your sound card and then returns it to the default setting when you are done.

Don't Overdrive Your Transceiver!

When you're setting up your rig for your first PSK31 transmission, the temptation is to adjust the output settings for "maximum smoke." This can be a serious mistake because overdriving your transceiver in PSK31 can result in a horrendous amount of splatter, which will suddenly make your PSK31 signal much wider than 31 Hz — and make you highly unpopular with operators on adjacent frequencies.

As you increase the transmit audio output from your sound card or multimode processor, watch the ALC indicator on your transceiver. The ALC is the automatic level control that governs the audio drive level. When you see your ALC display indicating that audio limiting is taking place, you are feeding too much audio to the transceiver. The goal is to achieve the desired RF output with little or no activation of the ALC.

Unfortunately, monitoring the ALC by itself is not always a sure bet. Many radios can be driven to full output without budging the ALC meter. You'd think that it would be smooth sailing from there, but a number of rigs become decidedly nonlinear when asked to provide SSB output beyond a certain level. (Sometimes this nonlinearity can begin at the 50% output level.) We can ignore the linearity issue to a certain extent with an SSB voice signal, but not with PSK31 because the immediate result, once again, is splatter.

So how can you tell if your PSK31 signal is really clean? Unless you have the means to monitor your RF output with an oscilloscope, the only way to check your signal is to ask someone to give you an evaluation on the air. The PSK31 programs that use a waterfall audio spectrum display can easily detect "dirty" signals. The splatter appears as rows of lines extending to the right and left of your primary signal. (Overdriven PSK31 signals may also have a harsh, clicking sound.)

If you are told that you are splattering, ask the other station to observe your signal as you slowly decrease the audio level from the sound card or processor. When you reach the point where the splatter disappears, you're all set. Don't worry if you discover that you can only generate a clean signal at, say, 50 W output. With PSK31 the performance differential between 50 W and 100 W is inconsequential.

SO WHAT'S OUT THERE?

What can you do with your new HF digital station? More than you might imagine. There are more than a dozen different HF digital modes on the air today, but only a handful are used extensively. Let's concentrate on the modes you're most likely to encounter.

RTTY

Good old radioteletype (RTTY) is still holding its ground on the HF bands. In fact, it is still the most popular digital mode for contests and DXpeditions (where groups of hams travel to rare DX Century Club locations to operate).

As with so many aspects of Amateur Radio, it is best to begin by listening. **Table 8-1** shows where to find digital mode activity on the various HF bands. There's almost always something happening on 20 meters, so try there first. Tune between 14.070 and 14.095 MHz and listen for the long, continuous *blee-blee-blee-blee* signals of RTTY. What you are hearing are the two alternating *mark* and *space* signals that RTTY uses in its binary *Baudot* coding scheme.

**Table 8-1
Popular HF Digital Frequencies**

Band (Meters)	Frequencies (MHz)
10	28.070–28.120
12	24.920–24.930
15	21.070–21.110
17	18.100–18.110
20	14.070–14.099
30	10.130–10.140
40	7.070–7.125
80	3.570–3.600
160	1.800–1.810

Depending on your transceiver and interface, you may use *audio frequency shift keying* (*AFSK*) or *frequency shift keying* (*FSK*) to transmit RTTY. The end result sounds the same on the air.

In AFSK, the audio tones for mark and space are generated by the computer and/or interface and fed into your transceiver's microphone input. For AFSK, make sure your transceiver is set for lower sideband (LSB). That is the RTTY convention.

Your rig may have an FSK mode (sometimes labeled DATA or RTTY). Many hams prefer operating this way because it allows them to use the transceiver's narrow IF filters to screen out interference. Just like on CW, with RTTY you can use 500 Hz or narrower filters to get rid of nearby signals. When you operate in FSK mode, your computer is not generating the mark and space tones. It is merely sending data pulses to the radio and the radio is creating its own mark and space signals. (This often requires a special connection to the transceiver. Consult your manual.)

A good time to observe RTTY activity is during a contest. There is at least one RTTY contest every month. See the sidebar "HF Digital Contesting."

With your software in the RTTY mode, tune across a RTTY signal. Your software likely includes some kind of tuning indicator to help you line up the mark and space signals correctly so that the software can start decoding. The tuning indicator may be part of a waterfall display. It might consist of two parallel lines that line up with mark and space, or it could even be a simulated oscilloscope display. Check your software manual and experiment with any variable settings. With some experience, you will be able to correlate the sound of the RTTY signal with the visual indicator and quickly tune in a new signal.

As you tune in the signal, you should see letters marching

good sound card can give you a substantial edge. This is particularly true if you are using a software-defined radio. Sound card performance is critical for this application.

In 2007 the ARRL undertook an evaluation of 11 common sound card models. The study was performed by Jonathan Taylor, K1RFD, and the results were published in the "Product Review" section of the May 2007 issue of *QST*. As you'd expect, the high-end sound cards came out on top, but think carefully before you reach for your wallet. *All* of the sound cards tested delivered adequate performance for digital mode applications. Don't buy more performance than you really need.

HF Digital Contesting

Some hams shun contesting because they assume they don't have the time or hardware necessary to win — and they are probably right. But winning is not the objective for most contesters. You enter a contest to do the best you can, to push yourself and your station to whatever limits you wish. The satisfaction at the end of a contest comes from the knowledge that you were part of the glorious frenzy, and that you gave it your best shot!

Contesting also has a practical benefit. If you're an award chaser, you can work many desirable stations during an active contest. During the ARRL RTTY Roundup, for example, some hams have worked enough international stations to earn a RTTY DXCC. Jump into the North American RTTY QSO Party and you stand a good chance of earning your Worked All States in a single weekend.

Contest Software

No one says you have to use software to keep track of your contest contacts, but it certainly makes life easier! One of the fundamental elements of any contest program is the ability to check for duplicate contacts or dupes. Working the same station that you just worked an hour ago is not only embarrassing, it is a waste of time. The better contest programs feature immediate dupe checking. When you enter the call sign in the logging window, the software instantly checks your log and warns you if the contact qualifies as a dupe. The more sophisticated programs "know" the rules of all the popular digital contests and they can quickly determine whether a contact is truly a dupe under the rules of the contest in question. Some contests, for example, allow you to work stations only once, regardless of the band. Other contests will allow you to work stations once per band.

A good software package will also help you track multipliers. It will display a list of multipliers you've worked, or show the ones you still need to find.

One of the most widely used contest software packages in the HF digital world is *Writelog* for *Windows*, which you'll find at www.writelog.com. *Writelog* has built-in RTTY and PSK31 modules. Another digital contest favorite is the free *N1MM Logger* at pages.cthome.net/n1mm/. *N1MM* by itself does not have RTTY capability, but you can add this functionality by installing the free *MMTTY* RTTY application, which is available at mmhamsoft. amateur-radio.ca/mmtty/.

For more information about contesting, see Chapter 7.

Digital Contest Calendar

See *QST* or the *National Contest Journal* for complete rules and to check dates.

Month	Date	Contest
January	1st	SARTG New Year's RTTY Contest
January	First full weekend	ARRL RTTY Roundup
January	Fourth full weekend	BARTG RTTY Sprint
February	First full weekend	NW QRP Club Digital Contest
February	Second weekend	CQ WW RTTY WPX Contest
February	Last full weekend	North American RTTY QSO Party
March	Second full weekend	*NCJ* RTTY Sprint
March	Third weekend	BARTG Spring RTTY Contest
April	First full weekend	EA WW RTTY Contest
April	Third weekend	TARA PSK31 Rumble
April	Fourth full weekend	SP DX RTTY Contest
May	First full weekend	ARI International DX Contest
May	Second weekend	VOLTA WW RTTY Contest
July	Third weekend	North American RTTY QSO Party
July	Fourth full weekend	Russian WW RTTY Contest
August	Third weekend	SARTG WW RTTY Contest
August	Last full weekend	SCC RTTY Championship
September	First full weekend	CCCC PSK31 Contest
September	Last full weekend	CQ WW RTTY Contest
October	First full weekend	TARA PSK31 Rumble
October	First full weekend	DARC International Hellschreiber Contest
October	Second Thursday	Internet RTTY Sprints
October	Second full weekend	*NCJ* RTTY Sprint
October	Third weekend	JARTS WW RTTY Contest
November	Second weekend	WAE RTTY Contest
December	First full weekend	TARA RTTY Sprints
December	Second weekend	OK RTTY DX Contest

Writelog contest software.

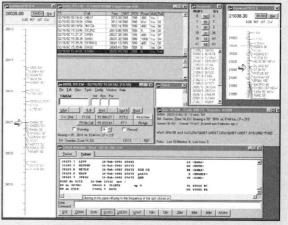

N1MM Logger.

across your screen. If you've stumbled across a contest, you may see something like this…

AA5AU 599 CT CT 010 010 DE WB8IMY K

In this instance, AA5AU is receiving a 599 signal report from WB8IMY in the state of Connecticut. This is also WB8IMY's tenth contact during the contest (that's the repeated "010"). The letter "K" means "over to you."

Most contest operators create "canned" messages, called *macros*, in their software. A macro can be set up to place your radio in the transmit mode, send a string of text, then return the radio to receive. Macros eliminate the need to type the same thing over and over, which comes in handy in a contest where you may be making hundreds of contacts.

Let's assume that you have a contest CQ stored in a macro right now. Tap the appropriate key (or click your mouse on the designated button) and your RTTY macro will do the rest…

(The radio enters the transmit mode)

CQ CONTEST CQ CONTEST DE WB8IMY WB8IMY K

(The radio returns to receive)

Of course hams still enjoy chatting on RTTY, just like they do on SSB or CW, and you may see them exchanging names, locations, antenna descriptions and other items of interest. RTTY DXing is popular too, and so you may run across "pile-ups" on DX stations. If the DX station is rare, the QSOs may just be rapid-fire exchange of signal reports and call signs, just like on voice or CW.

PSK31

PSK31 is the most widely used HF digital communications mode on the HF bands today. Most PSK31 activity involves casual conversation, although there are a few PSK31 contests as well.

The "PSK" stands for Phase Shift Keying, the modulation method that is used to generate the signal; "31" is the bit rate. Where RTTY uses two specific frequencies to communicate the binary data, PSK31 does the same thing by creating an audio signal that shifts its phase 180° in sync with the 31.25 bit-per-second data stream. A 0 bit in the data stream generates an audio phase shift, but a 1 does not. The technique of using phase shifts (and the lack thereof) to represent binary data is known as Binary Phase-Shift Keying, or *BPSK*. If you apply a BPSK audio signal to an SSB transceiver, you end up with BPSK modulated RF. At this data rate the resulting PSK31 RF signal is only about 50 Hz wide, which is actually narrower than the average CW signal.

With such a narrow bandwidth, PSK31 makes the most of a very small amount of spectrum. Transmit power is highly concentrated, meaning that you don't need a lot of power to communicate over great distances. (Most PSK31 operators use less than 50 W output.) At the receiving end, the PSK31 software uses digital signal processing to detect the phase changes, even in very weak signals. The result is that PSK31 rivals or exceeds the weak-signal performance of CW.

Its terrific performance not withstanding, PSK31 will not always provide 100% copy; it is as vulnerable to interference as any digital mode. And there are times, during a geomagnetic storm for example, when ionospheric propagation will cause slight changes in the frequency of the signal you're trying to copy. (When you are trying to receive a narrow-bandwidth, phase-shifting signal, frequency stability is very important.) This effect is almost always confined to the polar regions and it shows up as very rapid flutter, which is deadly to PSK31. The good news is that these events are usually short-lived.

If you are operating your transceiver in SSB without using narrow IF or audio-frequency filtering, the bandwidth of the receive audio that you're dumping to your sound card is about 2000 to 3000 Hz. With a bandwidth of only about 50 Hz, a lot of PSK31 signals can squeeze into that spectrum. Your software acts like an audio spectrum analyzer, sweeping through the received audio from 100 to 3000 Hz and showing you the results in a waterfall display that continuously scrolls from top to bottom. What you see on your monitor are vertical lines of various colors that indicate every signal the software can detect. Bright yellow lines represent strong signals while blue lines indicate weaker signals.

Most of the PSK31 signals on 20 meters are clustered around 14.070 MHz. You'll also find PSK31 activity on 3.580, 7.070 and 21.070 MHz. PSK31 signals have a distinctive sound unlike any digital mode you've heard on the ham bands. You won't find PSK31 by listening for the *deedle-deedle* of a RTTY signal, and PSK31 doesn't "chirp" like the TOR modes. PSK31 signals *warble*.

Start by putting your radio in the USB (upper sideband) mode and parking it on or near a PSK31 frequency (tune until you see a number of lines in the waterfall). Do not touch your rig's VFO again. Just boot up your software and get ready to have fun.

It is not at all uncommon to see several strong signals (the audible ones) interspersed with wispy blue ghosts of very weak "silent" signals. Click on a few of these ghosts and you may be rewarded with text (not error free, but good enough to understand what is being discussed).

As you decode PSK31 signals, the results will be a conversation on your monitor…

Yes, John, I'm seeing perfect text on my screen, but I can barely hear your signal. PSK31 is amazing! KF6I DE WB8IMY K

I know what you mean, Steve. You are also weak on my end, but 100% copy. WB8IMY DE KF6I K

If you find a station calling CQ and you want to reply, don't worry about tuning your radio. The software will use your sound card to generate the PSK31 transmit signal at exactly the same audio frequency as the received signal. When applied to your radio, this audio signal will create an RF signal that is exactly where it needs to be.

Some PSK31 programs and processor software offer type-ahead buffers, which allow you to compose your response "off line" while you are reading the incoming text from the other station. Just type what you wish to send, then press the keyboard key or click on the software "button" to transmit.

MFSK16

An MFSK signal consists of 16 tones, sent one at a time at

Fig 8-4 — Hellschreiber signals are displayed in repeating lines that look like an old-fashioned dot-matrix printer.

15.625 baud, and they are spaced only 15.625 Hz apart. Each tone represents four binary bits of data. With a bandwidth of 316 Hz, the signal easily fits through a narrow CW filter. MFSK16 has a distinctive musical sound that some compare to an old-fashioned carnival calliope.

MFSK16 can be tricky to tune. You must place the cursor at exactly the right spot on the signal pattern in the waterfall display. It takes some skill and patience to tune MFSK, but the results are worth the effort. MFSK offers excellent weak-signal performance and is a conversational mode like PSK31. Listen for the "music" of MFSK just above the PSK31 frequencies.

Hellschreiber

Are the Hellschreiber modes really HF digital? Or, are they a hybrid of the analog and digital worlds? Check out **Fig 8-4**.

Some argue that the Hellschreiber modes are more closely related to facsimile since they display text on your computer screen in the form of images (not unlike the product of a fax machine). On the other hand, the elements of the Hellschreiber "image text" are transmitted using a strictly defined digital format rather than the various analog signals of true HF fax or SSTV.

The Hellschreiber concept itself is quite old, developed in the 1920s by Rudolf Hell. Hellschreiber was, in fact, the first successful direct printing text transmission system. The German Army used Hellschreiber for field communications in World War II, and the mode was in use for commercial landline service until about 1980.

As personal computers became ubiquitous tools in ham shacks throughout the world, interest in Hellschreiber as an HF mode increased. By the end of the 20th century amateurs had developed several sophisticated pieces of Hellschreiber software and had also expanded and improved the Hellschreiber system itself.

Like PSK31, the Hellschreiber modes are intended for live conversations. Most of the activity is found on 20 meters, typically between 14.076 and 14.080 MHz. As with other conversational modes, you simply type and send your text. The main difference involves what is actually seen on the receiving end.

Feld-Hell is the most popular Hellschreiber mode among HF digital experimenters. It has its roots in the original Hellschreiber format, adapted slightly for ham use. Each character of a Feld-Hell transmission is communicated as a series of dots, with the result looking a bit like the output from a dot-matrix printer. A key-down state is used to indicate the black area of text, and the key up state is used to indicate blank or white spaces. One hundred and fifty characters are transmitted every minute. Each character takes 400 ms to complete. Because there are 49 pixels per character, each pixel is 8.163 ms long.

Most multimode software packages offer Hellschreiber. You won't find many Hell signals on the air, but conversing in this mode is a unique experience. Listen for odd "scratchy" signals just above the PSK31 frequencies.

References

From Chapter 22 of the 2010 ARRL Handbook

RF Connectors and Transmission Lines

There are many different types of transmission lines and RF connectors for coaxial cable, but the three most common for amateur use are the UHF, Type N and BNC families. The type of connector used for a specific job depends on the size of the cable, the frequency of operation and the power levels involved. **Table 22.60** shows the characteristics of many popular transmission lines, while **Table 22.61** details coax connectors.

UHF Connectors

The so-called UHF connector (the series name is not related to frequency) is found on most HF and some VHF equipment. It is the only connector many hams will ever see on coaxial cable. PL-259 is another name for the UHF male, and the female is also known as the SO-239. These connectors are rated for full legal amateur power at HF. They are poor for UHF work because they do not present a constant impedance, so the UHF label is a misnomer. PL-259 connectors are designed to fit RG-8 and RG-11 size cable (0.405-inch OD). Adapters are available for use with smaller RG-58, RG-59 and RG-8X size cable. UHF connectors are not weatherproof.

Fig 22.19 shows how to install the solder type of PL-259 on RG-8 cable. Proper preparation of the cable end is the key to success. Follow these simple steps. Measure back about ¾-inch from the cable end and slightly score the outer jacket around its circumference. With a sharp knife, cut through the outer jacket, through the braid and through the dielectric — almost to the center conductor. Be careful not to score the center conductor. Cutting through all outer layers at once keeps the braid from separating. (Using a coax stripping tool with preset blade depth makes this and subsequent trimming steps much easier.)

Pull the severed outer jacket, braid and dielectric off the end of the cable as one piece. Inspect the area around the cut, looking for any strands of braid hanging loose and snip them off. There won't be any if your knife was sharp enough. Next, score the outer jacket about 5/16-inch back from the first cut. Cut through the jacket lightly; do not score the braid. This step takes practice. If you score the braid, start again. Remove the outer jacket.

Tin the exposed braid and center conductor, but apply the solder sparingly and avoid melting the dielectric. Slide the coupling ring onto the cable. Screw the connector body onto the cable. If you prepared the cable to the right dimensions, the center conductor will protrude through the center pin, the braid will show through the solder holes, and the body will actually thread onto the outer cable jacket. A very small amount of lubricant on the cable jacket will help the threading process.

Solder the braid through the solder holes. Solder through all four holes; poor connection

Fig 22.20 — Installing PL-259 plugs on RG-58 or RG-59 cable requires the use of UG-175 or UG-176 adapters, respectively. The adapter screws into the plug body using the threads of the connector that grip the jacket on larger cables. (*Courtesy Amphenol Electronic Components*)

83-1SP (PL-259) Plug with adapters (UG-176/U OR UG-175/U)

1. Cut end of cable even. Remove vinyl jacket ¾" - don't nick braid. Slide coupling ring and adapter on cable.

2. Fan braid slightly and fold back over cable.

3. Position adapter to dimension shown. Press braid down over body of adapter and trim to 3/8". Bare 5/8" of conductor. Tin exposed center conductor.

4. Screw the plug assembly on adapter. Solder braid to shell through solder holes. Solder conductor to contact sleeve.

5. Screw coupling ring on plug assembly.

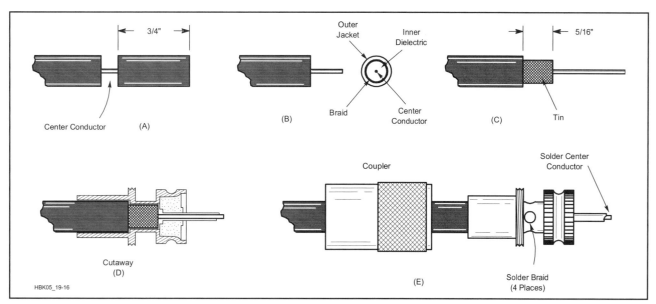

Fig 22.19 — The PL-259, or UHF, connector is almost universal for amateur HF work and is popular for equipment operating in the VHF range. Steps A through E are described in detail in the text.

to the braid is the most common form of PL-259 failure. A good connection between connector and braid is just as important as that between the center conductor and connector. Use a large soldering iron for this job. With practice, you'll learn how much heat to use. If you use too little heat, the solder will bead up, not really flowing onto the connector body. If you use too much heat, the dielectric will melt, letting the braid and center conductor touch. Most PL-259s are nickel plated, but silver-plated connectors are much easier to solder and only slightly more expensive.

Solder the center conductor to the center pin. The solder should flow on the inside, not the outside, of the center pin. If you wait until the connector body cools off from soldering the braid, you'll have less trouble with the dielectric melting. Trim the center conductor to be even with the end of the center pin. Use a small file to round the end, removing any solder that built up on the outer surface of the center pin. Use a sharp knife, very fine sandpaper or steel wool to remove any solder flux from the outer surface of the center pin. Screw the coupling ring onto the body, and you're finished.

Fig 22.20 shows how to install a PL-259 connector on RG-58 or RG-59 cable. An adapter is used for the smaller cable with standard RG-8 size PL-259s. (UG-175 for RG-58 and UG-176 for RG-59.) Prepare the cable as shown. Once the braid is prepared, screw the adapter into the PL-259 shell and finish the job as you would a PL-259 on RG-8 cable.

Fig 22.21 shows the instructions and dimensions for crimp-on UHF connectors that fit all common sizes of coaxial cable. While amateurs have been reluctant to adopt crimp-on connectors, the availability of good quality connectors and inexpensive crimping tools make crimp technology a good choice, even for connectors used outside. Soldering the center conductor to the connector tip is optional.

UHF connectors are not waterproof and must be waterproofed whether soldered or crimped as shown in the section of the **Safety** chapter on Antenna and Tower Safety.

BNC, N and F Connectors

The BNC connectors illustrated in **Fig 22.22** are popular for low power levels at VHF and UHF. They accept RG-58 and RG-59 cable, and are available for cable mounting in both male and female versions. Several different styles are available, so be sure to use the dimensions for the type you have. Follow the installation instructions carefully. If you prepare the cable to the wrong dimensions, the center pin will not seat properly with connectors of the opposite gender. Sharp scissors are a big help for trimming the braid evenly. Crimp-on BNC connectors are also available, with a large number of variations, including a twist-on version. A guide to installing these connectors is available on the CD-ROM accompanying this book.

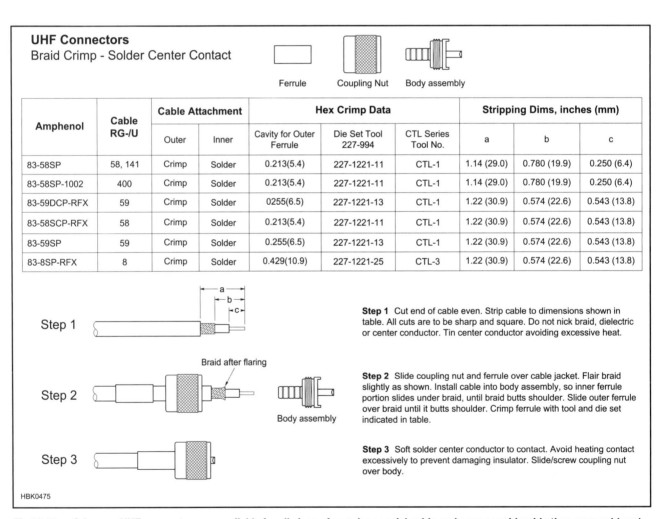

UHF Connectors
Braid Crimp - Solder Center Contact

Amphenol	Cable RG-/U	Cable Attachment		Hex Crimp Data			Stripping Dims, inches (mm)		
		Outer	Inner	Cavity for Outer Ferrule	Die Set Tool 227-994	CTL Series Tool No.	a	b	c
83-58SP	58, 141	Crimp	Solder	0.213(5.4)	227-1221-11	CTL-1	1.14 (29.0)	0.780 (19.9)	0.250 (6.4)
83-58SP-1002	400	Crimp	Solder	0.213(5.4)	227-1221-11	CTL-1	1.14 (29.0)	0.780 (19.9)	0.250 (6.4)
83-59DCP-RFX	59	Crimp	Solder	0255(6.5)	227-1221-13	CTL-1	1.22 (30.9)	0.574 (22.6)	0.543 (13.8)
83-58SCP-RFX	58	Crimp	Solder	0.213(5.4)	227-1221-11	CTL-1	1.22 (30.9)	0.574 (22.6)	0.543 (13.8)
83-59SP	59	Crimp	Solder	0.255(6.5)	227-1221-13	CTL-1	1.22 (30.9)	0.574 (22.6)	0.543 (13.8)
83-8SP-RFX	8	Crimp	Solder	0.429(10.9)	227-1221-25	CTL-3	1.22 (30.9)	0.574 (22.6)	0.543 (13.8)

Step 1 Cut end of cable even. Strip cable to dimensions shown in table. All cuts are to be sharp and square. Do not nick braid, dielectric or center conductor. Tin center conductor avoiding excessive heat.

Step 2 Slide coupling nut and ferrule over cable jacket. Flair braid slightly as shown. Install cable into body assembly, so inner ferrule portion slides under braid, until braid butts shoulder. Slide outer ferrule over braid until it butts shoulder. Crimp ferrule with tool and die set indicated in table.

Step 3 Soft solder center conductor to contact. Avoid heating contact excessively to prevent damaging insulator. Slide/screw coupling nut over body.

Fig 22.21 — Crimp-on UHF connectors are available for all sizes of popular coaxial cable and save considerable time over soldered connectors. The performance and reliability of these connectors is equivalent to soldered connectors, if crimped properly. (*Courtesy Amphenol Electronic Components*)

Table 22.60

Nominal Characteristics of Commonly Used Transmission Lines

RG or Type	Part Number	Nom. Z_0 Ω	VF %	Cap. pF/ft	Cent. Cond. AWG	Diel. Type	Shield Type	Jacket Matl	OD inches	Max V (RMS)	Matched Loss (dB/100') 1 MHz	10	100	1000
RG-6	Belden 1694A	75	82	16.2	#18 Solid BC	FPE	FC	P1	0.275	600	0.2	.7	1.8	5.9
RG-6	Belden 8215	75	66	20.5	#21 Solid CCS	PE	D	PE	0.332	2700	0.4	0.8	2.7	9.8
RG-8	Belden 7810A	50	86	23.0	#10 Solid BC	FPE	FC	PE	0.405	600	0.1	0.4	1.2	4.0
RG-8	TMS LMR400	50	85	23.9	#10 Solid CCA	FPE	FC	PE	0.405	600	0.1	0.4	1.3	4.1
RG-8	Belden 9913	50	84	24.6	#10 Solid BC	ASPE	FC	P1	0.405	600	0.1	0.4	1.3	4.5
RG-8	CXP1318FX	50	84	24.0	#10 Flex BC	FPE	FC	P2N	0.405	600	0.1	0.4	1.3	4.5
RG-8	Belden 9913F7	50	83	24.6	#11 Flex BC	FPE	FC	P1	0.405	600	0.2	0.6	1.5	4.8
RG-8	Belden 9914	50	82	24.8	#10 Solid BC	FPE	FC	P1	0.405	600	0.2	0.5	1.5	4.8
RG-8	TMS LMR400UF	50	85	23.9	#10 Flex BC	FPE	FC	PE	0.405	600	0.1	0.4	1.4	4.9
RG-8	DRF-BF	50	84	24.5	#9.5 Flex BC	FPE	FC	PE	0.405	600	0.1	0.4	1.6	5.2
RG-8	WM CQ106	50	84	24.5	#9.5 Flex BC	FPE	FC	P2N	0.405	600	0.2	0.6	1.8	5.3
RG-8	CXP008	50	78	26.0	#13 Flex BC	FPE	S	P1	0.405	600	0.1	0.5	1.8	7.1
RG-8	Belden 8237	52	66	29.5	#13 Flex BC	PE	S	P1	0.405	3700	0.2	0.6	1.9	7.4
RG-8X	Belden 7808A	50	86	23.5	#15 Solid BC	FPE	FC	PE	0.240	600	0.2	0.7	2.3	7.4
RG-8X	TMS LMR240	50	84	24.2	#15 Solid BC	FPE	FC	PE	0.242	300	0.2	0.8	2.5	8.0
RG-8X	WM CQ118	50	82	25.0	#16 Flex BC	FPE	FC	P2N	0.242	300	0.3	0.9	2.8	8.4
RG-8X	TMS LMR240UF	50	84	24.2	#15 Flex BC	FPE	FC	PE	0.242	300	0.2	0.8	2.8	9.6
RG-8X	Belden 9258	50	82	24.8	#16 Flex BC	FPE	S	P1	0.242	600	0.3	0.9	3.1	11.2
RG-8X	CXP08XB	50	80	25.3	#16 Flex BC	FPE	S	P1	0.242	300	0.3	0.9	3.1	14.0
RG-9	Belden 8242	51	66	30.0	#13 Flex SPC	PE	SCBC	P2N	0.420	5000	0.2	0.6	2.1	8.2
RG-11	Belden 8213	75	84	16.1	#14 Solid BC	FPE	S	PE	0.405	600	0.2	0.4	1.3	5.2
RG-11	Belden 8238	75	66	20.5	#18 Flex TC	PE	S	P1	0.405	600	0.2	0.7	2.0	7.1
RG-58	Belden 7807A	50	85	23.7	#18 Solid BC	FPE	FC	PE	0.195	300	0.3	1.0	3.0	9.7
RG-58	TMS LMR200	50	83	24.5	#17 Solid BC	FPE	FC	PE	0.195	300	0.3	1.0	3.2	10.5
RG-58	WM CQ124	52	66	28.5	#20 Solid BC	PE	S	PE	0.195	1400	0.4	1.3	4.3	14.3
RG-58	Belden 8240	52	66	28.5	#20 Solid BC	PE	S	P1	0.193	1900	0.3	1.1	3.8	14.5
RG-58A	Belden 8219	53	73	26.5	#20 Flex TC	FPE	S	P1	0.195	300	0.4	1.3	4.5	18.1
RG-58C	Belden 8262	50	66	30.8	#20 Flex TC	PE	S	P2N	0.195	1400	0.4	1.4	4.9	21.5
RG-58A	Belden 8259	50	66	30.8	#20 Flex TC	PE	S	P1	0.192	1900	0.4	1.5	5.4	22.8
RG-59	Belden 1426A	75	83	16.3	#20 Solid BC	FPE	S	P1	0.242	300	0.3	0.9	2.6	8.5
RG-59	CXP 0815	75	82	16.2	#20 Solid BC	FPE	S	P1	0.232	300	0.5	0.9	2.2	9.1
RG-59	Belden 8212	75	78	17.3	#20 Solid CCS	FPE	S	P1	0.242	300	0.6	1.0	3.0	10.9
RG-59	Belden 8241	75	66	20.4	#23 Solid CCS	PE	S	P1	0.242	1700	0.6	1.1	3.4	12.0
RG-62A	Belden 9269	93	84	13.5	#22 Solid CCS	ASPE	S	P1	0.240	750	0.3	0.9	2.7	8.7
RG-62B	Belden 8255	93	84	13.5	#24 FCC CCS	ASPE	S	P2N	0.242	750	0.3	0.9	2.9	11.0
RG-63B	Belden 9857	125	84	9.7	#22 Solid CCS	ASPE	S	P2N	0.405	750	0.2	0.5	1.5	5.8
RG-142	CXP 183242	50	69.5	29.4	#19 Solid SCCS	TFE	D	FEP	0.195	1900	0.3	1.1	3.8	12.8
RG-142B	Belden 83242	50	69.5	29.0	#19 Solid SCCS	TFE	D	TFE	0.195	1400	0.3	1.1	3.9	13.5
RG-174	Belden 7805R	50	73.5	26.2	#25 Solid BC	FPE	FC	P1	0.110	300	0.6	2.0	6.5	21.3
RG-174	Belden 8216	50	66	30.8	#26 Flex CCS	PE	S	P1	0.110	1100	1.9	3.3	8.4	34.0
RG-213	Belden 8267	50	66	30.8	#13 Flex BC	PE	S	P2N	0.405	3700	0.2	0.6	1.9	8.0
RG-213	CXP213	50	66	30.8	#13 Flex BC	PE	S	P2N	0.405	600	0.2	0.6	2.0	8.2
RG-214	Belden 8268	50	66	30.8	#13 Flex SPC	PE	D	P2N	0.425	3700	0.2	0.6	1.9	8.0
RG-216	Belden 9850	75	66	20.5	#18 Flex TC	PE	D	P2N	0.425	3700	0.2	0.7	2.0	7.1
RG-217	WM CQ217F	50	66	30.8	#10 Flex BC	PE	D	PE	0.545	7000	0.1	0.4	1.4	5.2
RG-217	M17/78-RG217	50	66	30.8	#10 Solid BC	PE	D	P2N	0.545	7000	0.1	0.4	1.4	5.2
RG-218	M17/79-RG218	50	66	29.5	#4.5 Solid BC	PE	S	P2N	0.870	11000	0.1	0.2	0.8	3.4
RG-223	Belden 9273	50	66	30.8	#19 Solid SPC	PE	D	P2N	0.212	1400	0.4	1.2	4.1	14.5
RG-303	Belden 84303	50	69.5	29.0	#18 Solid SCCS	TFE	S	TFE	0.170	1400	0.3	1.1	3.9	13.5
RG-316	CXP TJ1316	50	69.5	29.4	#26 Flex SCCS	TFE	S	FEP	0.098	1200	1.2	2.7	8.0	26.1
RG-316	Belden 84316	50	69.5	29.0	#26 Flex SCCS	TFE	S	FEP	0.096	900	1.2	2.7	8.3	29.0
RG-393	M17/127-RG393	50	69.5	29.4	#12 Flex SPC	TFE	D	FEP	0.390	5000	0.2	0.5	1.7	6.1
RG-400	M17/128-RG400	50	69.5	29.4	#20 Flex SPC	TFE	D	FEP	0.195	1400	0.4	1.1	3.9	13.2
LMR500	TMS LMR500UF	50	85	23.9	#7 Flex BC	FPE	FC	PE	0.500	2500	0.1	0.4	1.2	4.0
LMR500	TMS LMR500	50	85	23.9	#7 Solid CCA	FPE	FC	PE	0.500	2500	0.1	0.3	0.9	3.3
LMR600	TMS LMR600	50	86	23.4	#5.5 Solid CCA	FPE	FC	PE	0.590	4000	0.1	0.2	0.8	2.7
LMR600	TMS LMR600UF	50	86	23.4	#5.5 Flex BC	FPE	FC	PE	0.590	4000	0.1	0.2	0.8	2.7
LMR1200	TMS LMR1200	50	88	23.1	#0 Copper Tube	FPE	FC	PE	1.200	4500	0.04	0.1	0.4	1.3
Hardline														
1/2"	CATV Hardline	50	81	25.0	#5.5 BC	FPE	SM	none	0.500	2500	0.05	0.2	0.8	3.2
1/2"	CATV Hardline	75	81	16.7	#11.5 BC	FPE	SM	none	0.500	2500	0.1	0.2	0.8	3.2
7/8"	CATV Hardline	50	81	25.0	#1 BC	FPE	SM	none	0.875	4000	0.03	0.1	0.6	2.9
7/8"	CATV Hardline	75	81	16.7	#5.5 BC	FPE	SM	none	0.875	4000	0.03	0.1	0.6	2.9
LDF4-50A	Heliax – ½"	50	88	25.9	#5 Solid BC	FPE	CC	PE	0.630	1400	0.05	0.2	0.6	2.4
LDF5-50A	Heliax – ⅞"	50	88	25.9	0.355" BC	FPE	CC	PE	1.090	2100	0.03	0.10	0.4	1.3
LDF6-50A	Heliax – 1¼"	50	88	25.9	0.516" BC	FPE	CC	PE	1.550	3200	0.02	0.08	0.3	1.1
Parallel Lines														
TV Twinlead (Belden 9085)		300	80	4.5	#22 Flex CCS	PE	none	P1	0.400	**	0.1	0.3	1.4	5.9
Twinlead (Belden 8225)		300	80	4.4	#20 Flex BC	PE	none	P1	0.400	8000	0.1	0.2	1.1	4.8
Generic Window Line		405	91	2.5	#18 Solid CCS	PE	none	P1	1.000	10000	0.02	0.08	0.3	1.1
WM CQ 554		420	91	2.7	#14 Flex CCS	PE	none	P1	1.000	10000	0.02	0.08	0.3	1.1
WM CQ 552		440	91	2.5	#16 Flex CCS	PE	none	P1	1.000	10000	0.02	0.08	0.3	1.1
WM CQ 553		450	91	2.5	#18 Flex CCS	PE	none	P1	1.000	10000	0.02	0.08	0.3	1.1
WM CQ 551		450	91	2.5	#18 Solid CCS	PE	none	P1	1.000	10000	0.02	0.08	0.3	1.1
Open-Wire Line		600	92	1.1	#12 BC	none	none	none	**	12000	0.02	0.06	0.2	—

Approximate Power Handling Capability (1:1 SWR, 40°C Ambient):

	1.8 MHz	7	14	30	50	150	220	450	1 GHz
RG-58 Style	1350	700	500	350	250	150	120	100	50
RG-59 Style	2300	1100	800	550	400	250	200	130	90
RG-8X Style	1830	840	560	360	270	145	115	80	50
RG-8/213 Style	5900	3000	2000	1500	1000	600	500	350	250
RG-217 Style	20000	9200	6100	3900	2900	1500	1200	800	500
LDF4-50A	38000	18000	13000	8200	6200	3400	2800	1900	1200
LDF5-50A	67000	32000	22000	14000	11000	5900	4800	3200	2100
LMR500	18000	9200	6500	4400	3400	1900	1600	1100	700
LMR1200	52000	26000	19000	13000	10000	5500	4500	3000	2000

Legend:

**	Not Available or varies
ASPE	Air Spaced Polyethylene
BC	Bare Copper
CC	Corrugated Copper
CCA	Copper Cover Aluminum
CCS	Copper Covered Steel
CXP	Cable X-Perts, Inc.
D	Double Copper Braids
DRF	Davis RF
FC	Foil + Tinned Copper Braid
FEP	Teflon ® Type IX
Flex	Flexible Stranded Wire
FPE	Foamed Polyethylene
Heliax	Andrew Corp Heliax
N	Non-Contaminating
P1	PVC, Class 1
P2	PVC, Class 2
PE	Polyethylene
S	Single Braided Shield
SC	Silver Coated Braid
SCCS	Silver Plated Copper Coated Steel
SM	Smooth Aluminum
SPC	Silver Plated Copper
TC	Tinned Copper
TFE	Teflon®
TMS	Times Microwave Systems
UF	Ultra Flex
WM	Wireman

Fig 22.22 — BNC connectors are common on VHF and UHF equipment at low power levels. (*Courtesy Amphenol Electronic Components*)

BNC CONNECTORS

Standard Clamp

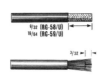

1. Cut cable even. Strip jacket. Fray braid and strip dielectric. **Don't nick braid or center conductor.** Tin center conductor.

2. Taper braid. Slide nut, washer, gasket and clamp over braid. Clamp inner shoulder should fit squarely against end of jacket.

3. With clamp in place, comb out braid, fold back smooth as shown. Trim center conductor.

4. Solder contact on conductor through solder hole. Contact should butt against dielectric. Remove excess solder from outside of contact. Avoid excess heat to prevent swollen dielectric which would interfere with connector body.

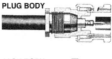

5. Push assembly into body. Screw nut into body with wrench until tight. **Don't rotate body on cable to tighten.**

Improved Clamp

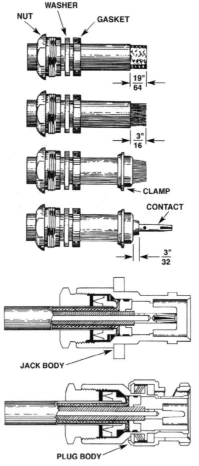

Follow 1, 2, 3 and 4 in BNC connectors (standard clamp) except as noted. Strip cable as shown. Slide gasket on cable *with groove facing clamp*. Slide clamp *with sharp edge facing gasket*. Clamp *should* cut gasket to seal properly.

C. C. Clamp

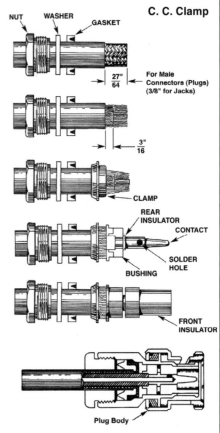

1. Follow steps 1, 2, and 3 as outlined for the standard-clamp BNC connector.

2. Slide on bushing, rear insulator and contact. The parts must butt securely against each other, as shown.

3. Solder the center conductor to the contact. Remove flux and excess solder.

4. Slide the front insulator over the contact, making sure it butts against the contact shoulder.

5. Insert the prepared cable end into the connector body and tighten the nut. Make sure the sharp edge of the clamp seats properly in the gasket.

Table 22.61

Coaxial Cable Connectors

UHF Connectors

Military No.	Style	Cable RG- or Description
PL-259	Str (m)	8, 9, 11, 13, 63, 87, 149, 213, 214, 216, 225
UG-111	Str (m)	59, 62, 71, 140, 210
SO-239	Pnl (f)	Std, mica/phenolic insulation
UG-266	Blkhd (f)	Rear mount, pressurized, copolymer of styrene ins.

Adapters

PL-258	Str (f/f)	Polystyrene ins.
UG-224, 363	Blkhd (f/f)	Polystyrene ins.
UG-646	Ang (f/m)	Polystyrene ins.
M-359A	Ang (m/f)	Polystyrene ins.
M-358	T (f/m/f)	Polystyrene ins.

Reducers

UG-175		55, 58, 141, 142 (except 55A)
UG-176		59, 62, 71, 140, 210

Family Characteristics:

All are nonweatherproof and have a nonconstant impedance. Frequency range: 0-500 MHz. Maximum voltage rating: 500 V (peak).

N Connectors

Military No.	Style	Cable RG-	Notes
UG-21	Str (m)	8, 9, 213, 214	50 Ω
UG-94A	Str (m)	11, 13, 149, 216	70 Ω
UG-536	Str (m)	58, 141, 142	50 Ω
UG-603	Str (m)	59, 62, 71, 140, 210	50 Ω
UG-23, B-E	Str (f)	8, 9, 87, 213, 214, 225	50 Ω
UG-602	Str (f)	59, 62, 71, 140, 210	—
UG-228B, D, E	Pnl (f)	8, 9, 87, 213, 214, 225	—
UG-1052	Pnl (f)	58, 141, 142	50 Ω
UG-593	Pnl (f)	59, 62, 71, 140, 210	50 Ω
UG-160A, B, D	Blkhd (f)	8, 9, 87, 213, 214, 225	50 Ω
UG-556	Blkhd (f)	58, 141, 142	50 Ω
UG-58, A	Pnl (f)		50 Ω
UG-997A	Ang (f)		50 Ω $^{11}/_{16}''$

Panel mount (f) with clearance above panel

M39012/04-	Blkhd (f)	Front mount hermetically sealed
UG-680	Blkhd (f)	Front mount pressurized

N Adapters

Military No.	Style	Notes
UG-29,A,B	Str (f/f)	50 Ω, TFE ins.
UG-57A,B	Str (m/m)	50 Ω, TFE ins.
UG-27A,B	Ang (f/m)	Mitre body
UG-212A	Ang (f/m)	Mitre body
UG-107A	T (f/m/f)	—
UG-28A	T (f/f/f)	—
UG-107B	T (f/m/f)	—

Family Characteristics:

N connectors with gaskets are weatherproof. RF leakage –90 dB min @ 3 GHz. Temperature limits: TFE: –67° to 390°F (–55° to 199°C). Insertion loss 0.15 dB max @ 10 GHz. Copolymer of styrene: –67° to 185°F (–55° to 85°C). Frequency range: 0-11 GHz. Maximum voltage rating: 1500 V P-P. Dielectric withstanding voltage 2500 V RMS. SWR (MIL-C-39012 cable connectors) 1.3 max 0-11 GHz.

BNC Connectors

Military No.	Style	Cable RG-	Notes
UG-88C	Str (m)	55, 58, 141, 142, 223, 400	

Military No.	Style	Cable RG-	Notes
UG-959	Str (m)	8, 9	
UG-260,A	Str (m)	59, 62, 71, 140, 210	Rexolite ins.
UG-262	Pnl (f)	59, 62, 71, 140, 210	Rexolite ins.
UG-262A	Pnl (f)	59, 62, 71, 140, 210	nwx, Rexolite ins.
UG-291	Pnl (f)	55, 58, 141, 142, 223, 400	
UG-291A	Pnl (f)	55, 58, 141, 142, 223, 400	nwx
UG-624	Blkhd (f)	59, 62, 71, 140, 210	Front mount Rexolite ins.
UG-1094A	Blkhd		Standard
UG-625B	Receptacle		
UG-625			

BNC Adapters

Military No.	Style	Notes
UG-491,A	Str (m/m)	
UG-491B	Str (m/m)	Berylium, outer contact
UG-914	Str (f/f)	
UG-306	Ang (f/m)	
UG-306A,B	Ang (f/m)	Berylium outer contact
UG-414,A	Pnl (f/f)	# 3-56 tapped flange holes
UG-306	Ang (f/m)	
UG-306A,B	Ang (f/m)	Berylium outer contact
UG-274	T (f/m/f)	
UG-274A,B	T (f/m/f)	Berylium outer contact

Family Characteristics:

Z = 50 Ω. Frequency range: 0-4 GHz w/low reflection; usable to 11 GHz. Voltage rating: 500 V P-P. Dielectric withstanding voltage 500 V RMS. SWR: 1.3 max 0-4 GHz. RF leakage –55 dB min @ 3 GHz. Insertion loss: 0.2 dB max @ 3 GHz. Temperature limits: TFE: –67° to 390°F (–55° to 199°C); Rexolite insulators: –67° to 185°F (–55° to 85°C). "Nwx" = not weatherproof.

HN Connectors

Military No.	Style	Cable RG-	Notes
UG-59A	Str (m)	8, 9, 213, 214	
UG-1214	Str (f)	8, 9, 87, 213, 214, 225	Captivated contact
UG-60A	Str (f)	8, 9, 213, 214	Copolymer of styrene ins.
UG-1215	Pnl (f)	8, 9, 87, 213, 214, 225	Captivated contact
UG-560	Pnl (f)		
UG-496	Pnl (f)		
UG-212C	Ang (f/m)		Berylium outer contact

Family Characteristics:

Connector Styles: Str = straight; Pnl = panel; Ang = Angle; Blkhd = bulkhead. Z = 50 Ω. Frequency range = 0-4 GHz. Maximum voltage rating = 1500 V P-P. Dielectric withstanding voltage = 5000 V RMS SWR = 1.3. All HN series are weatherproof. Temperature limits: TFE: –67° to 390°F (–55° to 199°C); copolymer of styrene: –67° to 185°F (–55° to 85°C).

Cross-Family Adapters

Families	Description	Military No.
HN to BNC	HN-m/BNC-f	UG-309
N to BNC	N-m/BNC-f	UG-201,A
	N-f/BNC-m	UG-349,A
	N-m/BNC-m	UG-1034
N to UHF	N-m/UHF-f	UG-146
	N-f/UHF-m	UG-83,B
	N-m/UHF-m	UG-318
UHF to BNC	UHF-m/BNC-f	UG-273
	UHF-f/BNC-m	UG-255

Type N assembly instructions

HBK05_19-19

CLAMP TYPES

     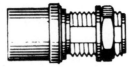

Nut Washer Gasket Clamp Male Contact Plug Body Female Contact Jack Body

Step 1

Step 2

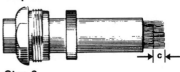

Step 3

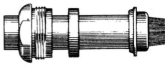

Step 4

Step 5

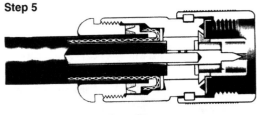

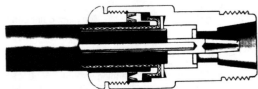

Amphenol Number	Connector Type	Cable RG-/U	Strip Dims., inches (mm)	
			a	c
82-61	N Plug	8, 9, 144, 165, 213, 214, 216, 225	0.359(9.1)	0.234(6.0)
82-62	N Panel Jack		0.312(7.9)	0.187(4.7)
82-63	N Jack	8, 9, 87A, 144, 165, 213, 214, 216, 225	0.281(7.1)	0.156(4.0)
82-67	N Bulkhead Jack			
82-202	N Plug	8, 9, 144, 165, 213, 214, 216, 225	0.359(9.1)	0.234(6.0)
82-202-1006	N Plug	Belden 9913	0.359(9.1)	0.234(6.0)
82-835	N Angle Plug	8, 9, 87A, 144, 165, 213, 214, 216, 225	0.281(7.1)	0.156(4.0)
18750	N Angle Plug	58, 141, 142	0.484(12.3)	0.234(5.9)
34025	N Plug		0.390(9.9)	0.203(5.2)
34525	N Plug	59, 62, 71, 140, 210	0.410(10.4)	0.230(5.8)
35025	N Jack	58, 141, 142	0.375(9.5)	0.187(4.7)
36500	N Jack	59, 62, 71, 140, 210	0.484(12.3)	0.200(5.1)

Step 1 Place nut and gasket, with "V" groove toward clamp, over cable and cut off jacket to dim. a.

Step 2 Comb out braid and fold out. Cut off cable dielectric to dim. c as shown.

Step 3 Pull braid wires forward and taper toward center conductor. Place clamp over braid and push back against cable jacket.

Step 4 Fold back braid wires as shown, trim braid to proper length and form over clamp as shown. Solder contact to center conductor.

Step 5 Insert cable and parts into connector body. Make sure sharp edge of clamp seats properly in gasket. Tighten nut.

US Customary Units and Conversion Factors

Linear Units

12 inches (in) = 1 foot (ft)
36 inches = 3 feet = 1 yard (yd)
1 rod = $5\frac{1}{2}$ yards = $16\frac{1}{2}$ feet
1 statute mile = 1760 yards = 5280 feet
1 nautical mile = 6076.11549 feet

Area

1 ft^2 = 144 in^2
1 yd^2 = 9 ft^2 = 1296 in^2
1 rod^2 = $30\frac{1}{4}$ yd^2
1 acre = 4840 yd^2 = 43,560 ft^2
1 acre = 160 rod^2
1 mile^2 = 640 acres

Volume

1 ft^3 = 1728 in^3
1 yd^3 = 27 ft^3

Liquid Volume Measure

1 fluid ounce (fl oz) = 8 fluid drams = 1.804 in
1 pint (pt) = 16 fl oz
1 quart (qt) = 2 pt = 32 fl oz = $57\frac{3}{4}$ in^3
1 gallon (gal) = 4 qt = 231 in^3
1 barrel = $31\frac{1}{2}$ gal

Dry Volume Measure

1 quart (qt) = 2 pints (pt) = 67.2 in^3
1 peck = 8 qt
1 bushel = 4 pecks = 2150.42 in^3

Avoirdupois Weight

1 dram (dr) = 27.343 grains (gr) or (gr a)
1 ounce (oz) = 437.5 gr
1 pound (lb) = 16 oz = 7000 gr
1 short ton = 2000 lb, 1 long ton = 2240 lb

Troy Weight

1 grain troy (gr t) = 1 grain avoirdupois
1 pennyweight (dwt) or (pwt) = 24 gr t
1 ounce troy (oz t) = 480 grains
1 lb t = 12 oz t = 5760 grains

Apothecaries' Weight

1 grain apothecaries' (gr ap)
    = 1 gr t = 1 gr
1 dram ap (dr ap) = 60 gr
1 oz ap = 1 oz t = 8 dr ap = 480 gr
1 lb ap = 1 lb t = 12 oz ap = 5760 gr

Conversion

Metric Unit = Metric Unit × US Unit

(Length)

mm	25.4	inch
cm	2.54	inch
cm	30.48	foot
m	0.3048	foot
m	0.9144	yard
km	1.609	mile
km	1.852	nautical mile

(Area)

mm^2	645.16	inch^2
cm^2	6.4516	in^2
cm^2	929.03	ft^2
m^2	0.0929	ft^2
cm^2	8361.3	yd^2
m^2	0.83613	yd^2
m^2	4047	acre
km^2	2.59	mi^2

(Mass)   **(Avoirdupois Weight)**

grams	0.0648	grains
g	28.349	oz
g	453.59	lb
kg	0.45359	lb
tonne	0.907	short ton
tonne	1.016	long ton

(Volume)

mm^3	16387.064	in^3
cm^3	16.387	in^3
m^3	0.028316	ft^3
m^3	0.764555	yd^3
ml	16.387	in^3
ml	29.57	fl oz
ml	473	pint
ml	946.333	quart
l	28.32	ft^3
l	0.9463	quart
l	3.785	gallon
l	1.101	dry quart
l	8.809	peck
l	35.238	bushel

(Mass)   **(Troy Weight)**

g	31.103	oz t
g	373.248	lb t

(Mass)   **(Apothecaries' Weight)**

g	3.387	dr ap
g	31.103	oz ap
g	373.248	lb ap

Multiply →
Metric Unit = Conversion Factor × US Customary Unit

← Divide
Metric Unit ÷ Conversion Factor = US Customary Unit

International System of Units (SI)—Metric Units

Prefix	Symbol	Multiplication Factor		
exe	E	10^{18}	=	1,000,000 000,000,000,000
peta	P	10^{15}	=	1,000 000,000,000,000
tera	T	10^{12}	=	1,000,000,000,000
giga	G	10^{9}	=	1,000,000,000
mega	M	10^{6}	=	1,000,000
kilo	k	10^{3}	=	1,000
hecto	h	10^{2}	=	100
deca	da	10^{1}	=	10
		10^{0}	=	1
deci	d	10^{-1}	=	0.1
centi	c	10^{-2}	=	0.01
milli	m	10^{-3}	=	0.001
micro	µ	10^{-6}	=	0.000001
nano	n	10^{-9}	=	0.000000001
pico	p	10^{-12}	=	0.000000000001
femto	f	10^{-15}	=	0.000000000000001
atto	a	10^{-18}	=	0.000000000000000001

Linear

1 meter (m) = 100 centimeters (cm) = 1000 millimeters (mm)

Area

$1 \ m^2 = 1 \times 10^4 \ cm^2 = 1 \times 10^6 \ mm^2$

Volume

$1 \ m^3 = 1 \times 10^6 \ cm^3 = 1 \times 10^9 \ mm^3$
1 liter (l) = 1000 cm^3 = 1×10^6 mm^3

Mass

1 kilogram (kg) = 1000 grams (g)
 (Approximately the mass of 1 liter of water)
1 metric ton (or tonne) = 1000 kg

Voltage-Power Conversion Table

Based on a 50-ohm system

Voltage			Power	
RMS	Peak-to-Peak	dBmV	Watts	dBm
0.01 µV	0.0283 µV	−100	2×10^{-18}	−147.0
0.02 µV	0.0566 µV	−93.98	8×10^{-18}	−141.0
0.04 µV	0.113 µV	−87.96	32×10^{-18}	−134.9
0.08 µV	0.226 µV	−81.94	128×10^{-18}	−128.9
0.1 µV	0.283 µV	−80.0	200×10^{-18}	−127.0
0.2 µV	0.566 µV	−73.98	800×10^{-18}	−121.0
0.4 µV	1.131 µV	−67.96	3.2×10^{-15}	−114.9
0.8 µV	2.236 µV	−61.94	12.8×10^{-15}	−108.9
1.0 µV	2.828 µV	−60.0	20.0×10^{15}	−107.0
2.0 µV	5.657 µV	−53.98	80.0×10^{-15}	−101.0
4.0 µV	11.31 µV	−47.96	320.0×10^{-15}	−94.95
8.0 µV	22.63 µV	−41.94	1.28×10^{-12}	−88.93
10.0 µV	28.28 µV	−40.00	2.0×10^{-12}	−86.99
20.0 µV	56.57 µV	−33.98	8.0×10^{-12}	−80.97
40.0 µV	113.1 µV	−27.96	32.0×10^{-12}	−74.95
80.0 µV	226.3 µV	−21.94	128.0×10^{-12}	−68.93
100.0 µV	282.8 µV	−20.0	200.0×10^{-12}	−66.99
200.0 µV	565.7 µV	−13.98	800.0×10^{-12}	−60.97
400.0 µV	1.131 mV	−7.959	3.2×10^{-9}	−54.95
800.0 µV	2.263 mV	−1.938	12.8×10^{-9}	−48.93
1.0 mV	2.828 mV	0.0	20.0×10^{-9}	−46.99
2.0 mV	5.657 mV	6.02	80.0×10^{-9}	−40.97
4.0 mV	11.31 mV	12.04	320×10^{-9}	−34.95
8.0 mV	22.63 mV	18.06	1.28 µW	−28.93
10.0 mV	28.28 mV	20.00	1 2.0 µW	−26.99
20.0 mV	56.57 mV	26.02	8.0 µW	−20.97
40.0 mV	113.1 mV	32.04	32.0 µW	−14.95
80.0 mV	226.3 mV	38.06	128.0 µW	−8.93
100.0 mV	282.8 mV	40.0	200.0 µW	−6.99
200.0 mV	565.7 mV	46.02	800.0 µW	−0.97
223.6 mV	632.4 mV	46.99	1.0 mW	0
400.0 mV	1.131 V	52.04	3.2 mW	5.05
800.0 mV	2.263 V	58.06	12.80 mW	11.07
1.0 V	2.828 V	60.0	20.0 mW	13.01
2.0 V	5.657 V	66.02	80.0 mW	19.03
4.0 V	11.31 V	72.04	320.0 mW	25.05
8.0 V	22.63 V	78.06	1.28 W	31.07
10.0 V	28.28 V	80.0	2.0 W	33.01
20.0 V	56.57 V	86.02	8.0 W	39.03
40.0 V	113.1 V	92.04	32.0 W	45.05
80.0 V	226.3 V	98.06	128.0 W	51.07
100.0 V	282.8 V	100.0	200.0 W	53.01
200.0 V	565.7 V	106.0	800.0 W	59.03
223.6 V	632.4 V	107.0	1,000.0 W	60.0
400.0 V	1,131.0 V	112.0	3,200.0 W	65.05
800.0 V	2,263.0 V	118.1	12,800.0 W	71.07
1000.0 V	2,828.0 V	120.0	20,000 W	73.01
2000.0 V	5,657.0 V	126.0	80,000 W	79.03
4000.0 V	11,310.0 V	132.0	320,000 W	85.05
8000.0 V	22,630.0 V	138.1	1.28 MW	91.07
10,000.0 V	28,280.0 V	140.0	2.0 MW	93.01

$$1/d = \frac{\lambda/2}{d} = \frac{300}{2f \times d}$$

$$1/d = \frac{300}{2f \times d} = \frac{300}{2 \times 7.2 \times \frac{0.081 \text{in}}{39.37 \text{in}/\text{m}}} = 10,126$$

$$\text{Length (ft)} = \frac{492 \times 0.95}{f \text{ (MHz)}} = \frac{468}{f \text{ (MHz)}}$$

$$\text{Length (ft)} = \frac{492 \times K}{f \text{ (MHz)}}$$

Reflection Coefficient, Attenuation, SWR and Return Loss

Reflection Coefficient (%)	Attenuation (dB)	Max SWR	Return Loss, dB	Reflection Coefficient (%)	Attenuation (dB)	Max SWR	Return Loss, dB
1.000	0.000434	1.020	40.00	45.351	1.0000	2.660	6.87
1.517	0.001000	1.031	36.38	48.000	1.1374	2.846	6.38
2.000	0.001738	1.041	33.98	50.000	1.2494	3.000	6.02
3.000	0.003910	1.062	30.46	52.000	1.3692	3.167	5.68
4.000	0.006954	1.083	27.96	54.042	1.5000	3.352	5.35
4.796	0.01000	1.101	26.38	56.234	1.6509	3.570	5.00
5.000	0.01087	1.105	26.02	58.000	1.7809	3.762	4.73
6.000	0.01566	1.128	24.44	60.000	1.9382	4.000	4.44
7.000	0.02133	1.151	23.10	60.749	2.0000	4.095	4.33
7.576	0.02500	1.164	22.41	63.000	2.1961	4.405	4.01
8.000	0.02788	1.174	21.94	66.156	2.5000	4.909	3.59
9.000	0.03532	1.198	20.92	66.667	2.5528	5.000	3.52
10.000	0.04365	1.222	20.00	70.627	3.0000	5.809	3.02
10.699	0.05000	1.240	19.41	70.711	3.0103	5.829	3.01
11.000	0.05287	1.247	19.17				
12.000	0.06299	1.273	18.42				
13.085	0.07500	1.301	17.66				
14.000	0.08597	1.326	17.08				
15.000	0.09883	1.353	16.48				
15.087	0.10000	1.355	16.43				
16.000	0.1126	1.381	15.92				
17.783	0.1396	1.433	15.00				
18.000	0.1430	1.439	14.89				
19.000	0.1597	1.469	14.42				
20.000	0.1773	1.500	13.98				
22.000	0.2155	1.564	13.15				
23.652	0.2500	1.620	12.52				
24.000	0.2577	1.632	12.40				
25.000	0.2803	1.667	12.04				
26.000	0.3040	1.703	11.70				
27.000	0.3287	1.740	11.37				
28.000	0.3546	1.778	11.06				
30.000	0.4096	1.857	10.46				
31.623	0.4576	1.925	10.00				
32.977	0.5000	1.984	9.64				
33.333	0.5115	2.000	9.54				
34.000	0.5335	2.030	9.37				
35.000	0.5675	2.077	9.12				
36.000	0.6028	2.125	8.87				
37.000	0.6394	2.175	8.64				
38.000	0.6773	2.226	8.40				
39.825	0.75000	2.324	8.00				
40.000	0.7572	2.333	7.96				
42.000	0.8428	2.448	7.54				
42.857	0.8814	2.500	7.36				
44.000	0.9345	2.571	7.13				

$$\text{Length (in)} = \frac{5904 \times K}{f\,(\text{MHz})}$$

where $\rho = 0.01 \times$ (reflection coefficient in %)

$$\frac{492}{50.1} = 9.82 \text{ ft}$$

where RL = return loss (dB)

$$\frac{(9.82 \text{ ft} \times 12 \text{ in./ft})}{0.5 \text{ in.}} = 235.7$$

where $X = A/10$ and $A =$ attenuation (dB)

$$\frac{492 \times 0.945}{50.1} = 9.28 \text{ ft}$$

Return loss (dB) = $-8.68589 \ln(\rho)$
where ln is the natural log (log to the base e)

Attenuation (dB) = $-4.34295 \ln(1-\rho^2)$
where ln is the natural log (log to the base e)

Abbreviations List

A
a—atto (prefix for 10^{-18})
A—ampere (unit of electrical current)
ac—alternating current
ACC—Affiliated Club Coordinator
ACSSB—amplitude-compandored single sideband
A/D—analog-to-digital
ADC—analog-to-digital converter
AF—audio frequency
AFC—automatic frequency control
AFSK—audio frequency-shift keying
AGC—automatic gain control
Ah—ampere hour
ALC—automatic level control
AM—amplitude modulation
AMRAD—Amateur Radio Research and Development Corporation
AMSAT—Radio Amateur Satellite Corporation
AMTOR—Amateur Teleprinting Over Radio
ANT—antenna
ARA—Amateur Radio Association
ARC—Amateur Radio Club
ARES—Amateur Radio Emergency Service
ARQ—Automatic repeat request
ARRL—American Radio Relay League
ARS—Amateur Radio Society (station)
ASCII—American National Standard Code for Information Interchange
ATV—amateur television
AVC—automatic volume control
AWG—American wire gauge
az-el—azimuth-elevation

B
B—bel; blower; susceptance; flux density, (inductors)
balun—balanced to unbalanced (transformer)
BC—broadcast
BCD—binary coded decimal
BCI—broadcast interference
Bd—baud (bids in single-channel binary data transmission)
BER—bit error rate
BFO—beat-frequency oscillator
bit—binary digit
bit/s—bits per second
BM—Bulletin Manager
BPF—band-pass filter
BPL—Brass Pounders League
BPL—Broadband over Power Line
BT—battery
BW—bandwidth
Bytes—Bytes

C
c—centi (prefix for 10^{-2})
C—coulomb (quantity of electric charge); capacitor
CAC—Contest Advisory Committee
CATVI—cable television interference
CB—Citizens Band (radio)
CBBS—computer bulletin-board service
CBMS—computer-based message system
CCITT—International Telegraph and Telephone Consultative Committee
CCTV—closed-circuit television
CCW—coherent CW
ccw—counterclockwise
CD—civil defense
cm—centimeter
CMOS—complementary-symmetry metal-oxide semiconductor
coax—coaxial cable
COR—carrier-operated relay
CP—code proficiency (award)
CPU—central processing unit
CRT—cathode ray tube
CT—center tap
CTCSS—continuous tone-coded squelch system
cw—clockwise
CW—continuous wave

D
d—deci (prefix for 10^{-1})
D—diode
da—deca (prefix for 10)
D/A—digital-to-analog
DAC—digital-to-analog converter
dB—decibel (0.1 bel)
dBi—decibels above (or below) isotropic antenna
dBm—decibels above (or below) 1 milliwatt
DBM—double balanced mixer
dBV—decibels above/below 1 V (in video, relative to 1 V P-P)
dBW—decibels above/below 1 W
dc—direct current
D-C—direct conversion
DDS—direct digital synthesis
DEC—District Emergency Coordinator
deg—degree
DET—detector
DF—direction finding; direction finder
DIP—dual in-line package
DMM—digital multimeter
DPDT—double-pole double-throw (switch)
DPSK—differential phase-shift keying
DPST—double-pole single-throw (switch)
DS—direct sequence (spread spectrum); display
DSB—double sideband
DSP—digital signal processing
DTMF—dual-tone multifrequency
DVM—digital voltmeter
DX—long distance; duplex
DXAC—DX Advisory Committee
DXCC—DX Century Club

E
e—base of natural logarithms (2.71828)
E—voltage
EA—ARRL Educational Advisor
EC—Emergency Coordinator
ECL—emitter-coupled logic
EHF—extremely high frequency (30-300 GHz)
EIA—Electronic Industries Alliance
EIRP—effective isotropic radiated power
ELF—extremely low frequency
ELT—emergency locator transmitter
EMC—electromagnetic compatibility
EME—earth-moon-earth (moonbounce)
EMF—electromotive force
EMI—electromagnetic interference
EMP—electromagnetic pulse
EOC—emergency operations center
EPROM—erasable programmable read only memory

F
f—femto (prefix for 10^{-15}); frequency
F—farad (capacitance unit); fuse
fax—facsimile
FCC—Federal Communications Commission
FD—Field Day
FEMA—Federal Emergency Management Agency
FET—field-effect transistor
FFT—fast Fourier transform
FL—filter
FM—frequency modulation
FMTV—frequency-modulated television
FSK—frequency-shift keying
FSTV—fast-scan (real-time) television
ft—foot (unit of length)

G
g—gram (unit of mass)
G—giga (prefix for 10^9); conductance
GaAs—gallium arsenide
GB—gigabytes
GDO—grid- or gate-dip oscillator
GHz—gigahertz (10^9 Hz)
GND—ground

H
h—hecto (prefix for 10^2)
H—henry (unit of inductance)
HF—high frequency (3-30 MHz)
HFO—high-frequency oscillator; heterodyne frequency oscillator
HPF—highest probable frequency; high-pass filter
Hz—hertz (unit of frequency, 1 cycle/s)

I
I—current, indicating lamp
IARU—International Amateur Radio Union
IC—integrated circuit
ID—identification; inside diameter
IEEE—Institute of Electrical and Electronics Engineers
IF—intermediate frequency

IMD—intermodulation distortion
in.—inch (unit of length)
in./s—inch per second (unit of velocity)
I/O—input/output
IRC—international reply coupon
ISB—independent sideband
ITF—Interference Task Force
ITU—International Telecommunication Union
ITU-T—ITU Telecommunication Standardization Bureau

J-K

j—operator for complex notation, as for reactive component of an impedance ($+j$ inductive; $-j$ capacitive)
J—joule (kg m^2/s^2) (energy or work unit); jack
JFET—junction field-effect transistor
k—kilo (prefix for 10^3); Boltzmann's constant (1.38x10^{-23} J/K)
K—kelvin (used without degree symbol) absolute temperature scale; relay
kB—kilobytes
kBd—1000 bauds
kbit—1024 bits
kbit/s—1024 bits per second
kbyte—1024 bytes
kg—kilogram
kHz—kilohertz
km—kilometer
kV—kilovolt
kW—kilowatt
kΩ—kilohm

L

l—liter (liquid volume)
L—lambert; inductor
lb—pound (force unit)
LC—inductance-capacitance
LCD—liquid crystal display
LED—light-emitting diode
LF—low frequency (30-300 kHz)
LHC—left-hand circular (polarization)
LO—local oscillator; Leadership Official
LP—log periodic
LS—loudspeaker
lsb—least significant bit
LSB—lower sideband
LSI—large-scale integration
LUF—lowest usable frequency

M

m—meter (length); milli (prefix for 10^{-3})
M—mega (prefix for 10^6); meter (instrument)
mA—milliampere
mAh—milliampere hour
MB—megabytes
MCP—multimode communications processor
MDS—Multipoint Distribution Service; minimum discernible (or detectable) signal
MF—medium frequency (300-3000 kHz)
mH—millihenry
MHz—megahertz

mi—mile, statute (unit of length)
mi/h (MPH)—mile per hour
mi/s—mile per second
mic—microphone
min—minute (time)
MIX—mixer
mm—millimeter
MOD—modulator
modem—modulator/demodulator
MOS—metal-oxide semiconductor
MOSFET—metal-oxide semiconductor field-effect transistor
MS—meteor scatter
ms—millisecond
m/s—meters per second
msb—most-significant bit
MSI—medium-scale integration
MSK—minimum-shift keying
MSO—message storage operation
MUF—maximum usable frequency
mV—millivolt
mW—milliwatt
MΩ—megohm

N

n—nano (prefix for 10^{-9}); number of turns (inductors)
NBFM—narrow-band frequency modulation
NC—no connection; normally closed
NCS—net-control station; National Communications System
nF—nanofarad
NF—noise figure
nH—nanohenry
NiCd—nickel cadmium
NM—Net Manager
NMOS—N-channel metal-oxide silicon
NO—normally open
NPN—negative-positive-negative (transistor)
NPRM—Notice of Proposed Rule Making (FCC)
ns—nanosecond
NTIA—National Telecommunications and Information Administration
NTS—National Traffic System

O

OBS—Official Bulletin Station
OD—outside diameter
OES—Official Emergency Station
OO—Official Observer
op amp—operational amplifier
ORS—Official Relay Station
OSC—oscillator
OSCAR—Orbiting Satellite Carrying Amateur Radio
OTC—Old Timer's Club
oz—ounce ($\frac{1}{16}$ pound)

P

p—pico (prefix for 10^{-12})
P—power; plug
PA—power amplifier
PACTOR—digital mode combining aspects of packet and AMTOR
PAM—pulse-amplitude modulation
PBS—packet bulletin-board system

PC—printed circuit
PD—power dissipation
PEP—peak envelope power
PEV—peak envelope voltage
pF—picofarad
pH—picohenry
PIC—Public Information Coordinator
PIN—positive-intrinsic-negative (semiconductor)
PIO—Public Information Officer
PIV—peak inverse voltage
PLC—Power Line Carrier
PLL—phase-locked loop
PM—phase modulation
PMOS—P-channel (metal-oxide semiconductor)
PNP—positive negative positive (transistor)
pot—potentiometer
P-P—peak to peak
ppd—postpaid
PROM—programmable read-only memory
PSAC—Public Service Advisory Committee
PSHR—Public Service Honor Roll
PTO—permeability-tuned oscillator
PTT—push to talk

Q-R

Q—figure of merit (tuned circuit); transistor
QRP—low power (less than 5-W output)
R—resistor
RACES—Radio Amateur Civil Emergency Service
RAM—random-access memory
RC—resistance-capacitance
R/C—radio control
RCC—Rag Chewer's Club
RDF—radio direction finding
RF—radio frequency
RFC—radio-frequency choke
RFI—radio-frequency interference
RHC—right-hand circular (polarization)
RIT—receiver incremental tuning
RLC—resistance-inductance-capacitance
RM—rule making (number assigned to petition)
r/min (RPM)—revolutions per minute
rms—root mean square
ROM—read-only memory
r/s—revolutions per second
RS—Radio Sputnik (Russian ham satellite)
RST—readability-strength-tone (CW signal report)
RTTY—radioteletype
RX—receiver, receiving

S

s—second (time)
S—siemens (unit of conductance); switch
SASE—self-addressed stamped envelope
SCF—switched capacitor filter
SCR—silicon controlled rectifier
SEC—Section Emergency Coordinator

SET—Simulated Emergency Test
SGL—State Government Liaison
SHF—super-high frequency (3-30 GHz)
SM—Section Manager; silver mica (capacitor)
S/N—signal-to-noise ratio
SPDT—single-pole double-throw (switch)
SPST—single-pole single-throw (switch)
SS—ARRL Sweepstakes; spread spectrum
SSB—single sideband
SSC—Special Service Club
SSI—small-scale integration
SSTV—slow-scan television
STM—Section Traffic Manager
SX—simplex
sync—synchronous, synchronizing
SWL—shortwave listener
SWR—standing-wave ratio

T
T—tera (prefix for 10^{12}); transformer
TA—ARRL Technical Advisor
TC—Technical Coordinator
TCC—Transcontinental Corps (NTS)
TCP/IP—Transmission Control Protocol/Internet Protocol
tfc—traffic
TNC—terminal node controller (packet radio)
TR—transmit/receive
TS—Technical Specialist
TTL—transistor-transistor logic
TTY—teletypewriter
TU—terminal unit
TV—television
TVI—television interference
TX—transmitter, transmitting

U
U—integrated circuit
UHF—ultra-high frequency (300 MHz to 3 GHz)
USB—upper sideband
UTC—Coordinated Universal Time (also abbreviated Z)
UV—ultraviolet

V
V—volt; vacuum tube
VCO—voltage-controlled oscillator
VCR—video cassette recorder
VDT—video-display terminal
VE—Volunteer Examiner
VEC—Volunteer Examiner Coordinator
VFO—variable-frequency oscillator
VHF—very-high frequency (30-300 MHz)
VLF—very-low frequency (3-30 kHz)
VLSI—very-large-scale integration
VMOS—V-topology metal-oxide-semiconductor
VOM—volt-ohmmeter
VOX—voice-operated switch
VR—voltage regulator
VSWR—voltage standing-wave ratio
VTVM—vacuum-tube voltmeter
VUCC—VHF/UHF Century Club
VXO—variable-frequency crystal oscillator

W
W—watt (kg $m^2 s^{-3}$), unit of power
WAC—Worked All Continents
WAS—Worked All States
WBFM—wide-band frequency modulation
WEFAX—weather facsimile
Wh—watthour
WPM—words per minute
WRC—World Radiocommunication Conference
WVDC—working voltage, direct current

X
X—reactance
XCVR—transceiver
XFMR—transformer
XIT—transmitter incremental tuning
XO—crystal oscillator
XTAL—crystal
XVTR—transverter

Y-Z
Y—crystal; admittance
YIG—yttrium iron garnet
Z—impedance; also see UTC

Numbers/Symbols
5BDXCC—Five-Band DXCC
5BWAC—Five-Band WAC
5BWAS—Five-Band WAS
6BWAC—Six-Band WAC
°—degree (plane angle)
°C—degree Celsius (temperature)
°F—degree Fahrenheit (temperature)
α—(alpha) angles; coefficients, attenuation constant, absorption factor, area, common-base forward current-transfer ratio of a bipolar transistor
β—(beta) angles; coefficients, phase constant, current gain of common-emitter transistor amplifiers
γ—(gamma) specific gravity, angles, electrical conductivity, propagation constant
Γ—(gamma) complex propagation constant
δ—(delta) increment or decrement; density; angles
Δ—(delta) increment or decrement determinant, permittivity
ε—(epsilon) dielectric constant; permittivity; electric intensity
ζ—(zeta) coordinates; coefficients
η—(eta) intrinsic impedance; efficiency; surface charge density; hysteresis; coordinate
θ—(theta) angular phase displacement; time constant; reluctance; angles
ι—(iota) unit vector
κ—(kappa) susceptibility; coupling coefficient
λ—(lambda) wavelength; attenuation constant
Λ—(lambda) permeance
μ—(mu) permeability; amplification factor; micro (prefix for 10^{-6})
μF—microfarad
μH—microhenry
μP—microprocessor
ξ—(xi) coordinates
π—(pi) ≈ 3.14159
ρ—(rho) resistivity; volume charge density; coordinates; reflection coefficient
σ—(sigma) surface charge density; complex propagation constant; electrical conductivity; leakage coefficient; deviation
Σ—(sigma) summation
τ—(tau) time constant; volume resistivity; time-phase displacement; transmission factor; density
ϕ—(phi) magnetic flux angles
Φ—(phi) summation
χ—(chi) electric susceptibility; angles
Ψ—(psi) dielectric flux; phase difference; coordinates; angles
ω—(omega) angular velocity $2\pi F$
Ω—(omega) resistance in ohms; solid angle

Antenna and Tower Safety

From Chapter 28 of the 2010 ARRL Handbook

By definition, all of the topics in this book are about radio telecommunications. For those communications, both receive and transmit antennas are required and those antennas need to be up in the air in order to work effectively. While antennas are covered elsewhere, this section will cover many of the topics associated with getting those antennas up there, along with related safety issues. A more complete treatment of techniques used to erect towers and antennas is available in the *ARRL Antenna Book* and in the *Up The Tower: The Complete Guide To Tower Construction* by Steve Morris, K7LXC.

28.2.1 Legal Considerations

Some antenna support structures fall under local building regulations as well as neighborhood restrictions. Many housing developments have Homeowner's Associations (HOAs) as well as Covenants, Conditions and Restrictions (CC&Rs) that may have a direct bearing on whether a tower or similar structure can be erected at all. This is a broad topic with many pitfalls. Detailed background on these topics is provided in *Antenna Zoning for the Radio Amateur* by Fred Hopengarten, K1VR, an attorney with extensive experience in towers and zoning. You may also want to contact one of the ARRL Field Organization's Volunteer Counsels.

Even without neighborhood issues, a building permit is likely to be required. With the proliferation of cellular and other commercial wireless devices and their attendant RF sites, many local governments now require that the structures meet local building codes. Again, K1VR's book is extremely helpful in sorting all this out. Building permit applications may also require Professional Engineer (PE) calculations and stamp (certification). The ARRL Field Organization's Volunteer Consulting Engineer program may be useful with the engineering side of your project.

28.2.2 Antenna Mounting Structures

TREES AND POLES

The original antenna supports were trees: if you've got them, use them. They're free and unregulated, so it couldn't be easier. Single-trunked varieties such as fir and pine trees are easier to use than the multi-trunked varieties. Multi-trunked trees are not impossible to use — they just require a lot more work. For dipoles or other types of wire antennas, plan for the tree to support an end of the wire; trying to install an inverted V or similar configuration is almost impossible due to all of the intervening branches.

Install an eye-screw with a pulley at the desired height, trim away enough limbs to create a "window" for the antenna through the branches and then attach a rope halyard to the antenna insulator. Here's a useful tip: Make the *halyard* a continuous loop as shown in **Fig 28.9**. Since it's almost always the antenna wire that breaks, a continuous halyard makes it easy to reattach the wire and insulator. With just a single halyard, if the antenna breaks, the tree will have to be climbed to reach the pulley, then reinstall and attach the line. If you're unable to climb the support tree, contact a local tree service.

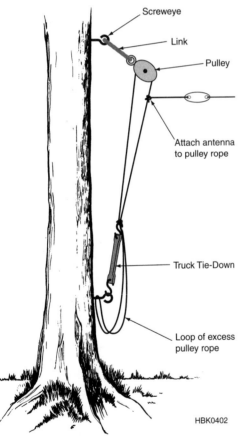

Fig 28.9 — Loop and halyard method of supporting wire antennas in trees. Should the antenna break, the continuous loop of rope allows antenna repair or replacement without climbing the tree.

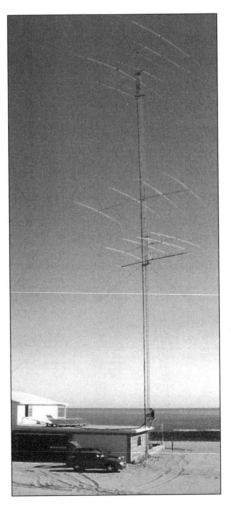

Fig 28.10 — A guyed tower with a good-sized load of antennas at XE2DV-W7ZR in Baja California, is shown at the left. At the right, the Trylon Titan self-supporting tower of W7WVF and N7YYG in Bandon, Oregon. (*K7LXC photos*)

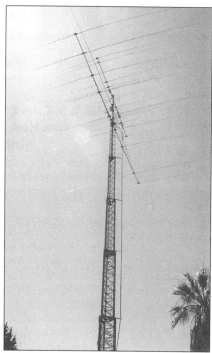

Fig 28.11 — The bottom of N6TV's crank-up tower is shown at left. The motor drive mechanism is on the left and a fishing net on the right catches and coils the feed lines and control cables as the tower is lowered. On the right, K6KR's fully loaded crank-up extended to its maximum height of 90 feet. (*K7LXC photos*)

Professional tree climbers are often willing to help out for a small fee.

Another way to get wires into trees is with some sort of launcher. Using a bow-and-arrow is a traditional method of shooting a fishing line over a tree to pull up a bigger line. There are now commercial products available that are easier to use and reach higher in the tree. For example, wrist-rocket slingshots and compressed-air launchers can reach heights of more than 200 feet!

Wooden utility poles offer a tree-related alternative but are not cheap, require special installation with a pole-setting truck, and there is no commercial antenna mounting hardware available for them. That makes them a poor choice for most installations.

TOWERS

The two most important parameters to consider when planning a tower installation are the maximum local *wind speed* and the proposed antenna *wind load*. Check with your local building department to find out what the maximum wind speed is for your area. Another source is a list of maximum wind speeds for all counties in the US from the TIA-222, *Structural Standard for Antenna Supporting Structures and Antennas*. This is an expensive professional publication so it's not for everyone, but the list is posted on the Champion Radio Products Web site under "Tech Notes." Tower capacities are generally specified in square feet of antenna load and antenna wind load specifications are provided by the antenna manufacturer.

Before beginning, learn and follow K7LXC's Prime Directive of tower construc-

tion — "DO what the manufacturer says." (And DON'T do what the manufacturer doesn't say to do.) Professional engineers have analyzed and calculated the proper specifications and conditions for tower structures and their environment. Taking any shortcuts or making different decisions will result in a less reliable installation.

Towers come in the two varieties shown in **Fig 28.10** — guyed and self-supporting. Guyed towers require a bigger footprint because the guys have to be anchored away from the tower — typically 80% of the tower height. Self-supporting towers need bigger bases to counteract the overturning moment and are more expensive than a guyed tower because there is more steel in them (the cost of a tower is largely determined by the cost of the steel).

The most popular guyed towers are the Rohn 25G and 45G. The 25G is a light-duty tower and the 45G is capable of carrying fairly big loads. The online Rohn catalog (see the References) has most of the information you'll need to plan an installation.

Self-supporting towers are made by several manufacturers and allow building a tower up to 100 feet or higher with a small footprint.

Fig 28.12 — The roof-mounted tower holding the AA2OW antenna system. (*AA2OW photo*)

Rohn, Trylon and AN Wireless are popular vendors. Another type of self-supporting tower is the *crank-up*, shown in **Fig 28.11**. Using a system of cables, pulleys and winches, crank-up towers can extend from 20 feet to over 100 feet high. These are moderately complex devices. The largest manufacturer of crank-up towers is US Towers.

Another simple and effective way to get an antenna up in the air is with a *roof-mounted tower*, seen in **Fig 28.12**. These are four-legged aluminum structures of heights from four to more than 20 feet. While they are designed to be lag-screwed directly into the roof trusses, it is often preferable to through-bolt them into a long 2×4 or 2×6 that acts a backing plate, straddling three to four roof trusses. In any case, if it is not clear how best to install the tower on the structure, have a roofing professional or engineer provide advice. Working on a roof-mounted tower also requires extra caution because of climbing on a roof while also working on a tower.

28.2.3 Tower Construction and Erection

THE BASE

Once all the necessary materials and the required approvals have been gathered, tower installation can begin. Let's assume you and your friends are going to install it. The first job is to construct the base. A base for a guyed tower can be hand-dug as can the guy anchors. For a self-supporting tower, renting an excavator of some sort will make it much easier to move the several cubic yards of dirt.

Next, some sort of rebar cage will be needed for the concrete. Guyed towers only require rudimentary rebar while a self-supporting tower will need a bigger, heavier and more elaborate cage. Consult the manufacturer's specifications for the exact materials and dimensions.

Typical tower concrete specs call for 3000 psi (minimum) compressive strength concrete and 28 days for a full cure. A local pre-mix concrete company can deliver it. Pouring the concrete is easiest if the concrete truck can back up to the hole. If that's not possible, a truck- or trailer-mounted line pump can pump it up to 400 feet at minimal expense if using a wheelbarrow is not possible or practical. Packaged concrete from the hardware store mixed manually may also be used. Quikrete Mix #1101 is rated at 2500 psi after seven days and 4000 psi after 28 days.

TOOLS

Once the base and anchors are finished and the concrete has cured, the tower can be constructed. There are several tools that will make the job easier. If the tower is a guyed tower, it can be erected either with a crane or a *gin-pole*. The gin-pole, shown in **Fig 28.13**, is a pipe that attaches to the leg of the tower and has a pulley at the top for the haul rope. Use the gin-pole to pull up one section at a time (see below).

Another useful tool for rigging and hoisting is the *carabiner*. Shown in **Fig 28.14** (A and B), carabiners are oval steel or aluminum snap-links popularized by mountain climb-

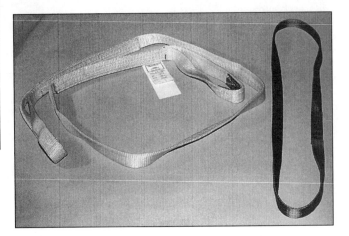

Fig 28.14 — (A) Oval mountain climbing type carabiners are ideal for tower workloads and attachments. The gates are spring loaded — the open gate is shown for illustration. (B) An open aluminum oval carabiner; a closed oval carabiner; an aluminum locking carabiner; a steel snaplink. (C) A heavy duty nylon sling on the left for big jobs and a lighter-duty loop sling on the right for everything else. (*K7LXC photos*)

Fig 28.13 — A gin-pole consists of a leg clamp fixture, a section of aluminum mast and a pulley. It is used to lift the tower section high enough to be safely lowered into place and attached. (Based on Rohn EF2545.)

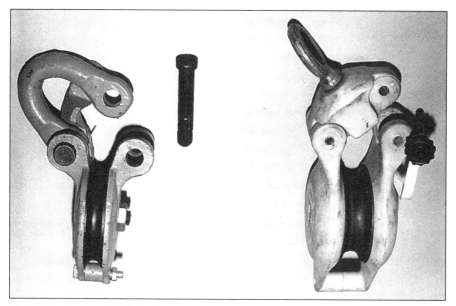

Fig 28.15 — Two snatch blocks; a steel version on the left and a lightweight high impact plastic one on the right in their open position. (*K7LXC photo*)

ers. They have spring-loaded gates and can be used for many tower tasks. For instance, there should be one at the end of the haul rope for easy and quick attachment to rotators, parts and tool buckets — virtually anything that needs to be raised or lowered. It can even act as a "third hand" on the tower.

Along with the carabineer, the *nylon loop sling* in Fig 28.14C can be wrapped around large or irregularly shaped objects such as antennas, masts or rotators and attached to ropes with carabiners. For a complex job, a professional will often climb with eight to ten slings and use every one!

A pulley or two will also make the job easier. At a minimum, one is needed for the haul rope at the top of the tower. A *snatch block* is also useful; this is a pulley whose top opens up to "snatch" (attach it to) the rope at any point. **Fig 28.15** shows two snatch-block pulleys used for tower work.

ROPES

Speaking of ropes, use a decent haul rope. Rope that is one-half inch diameter or larger affords a good grip for lifting and pulling. There are several choices of rope material. The best choice is a synthetic material such as nylon or Dacron. A typical twisted rope is fine for most applications. A synthetic rope with a braid over the twisted core is known as *braid-on-braid* or *kernmantle*. While it's more expensive than twisted ropes, the outer braid provides better abrasion resistance. The least expensive type of rope is polypropylene. It's a stiff rope that doesn't take a knot as well as other types but is reasonably durable and cheap. **Table 28.2** shows the safe working load ratings for common types of rope.

When doing tower work, being able to tie knots is required. Of all the knots, the *bowline* is the one to know for tower work. The old "rabbit comes up through the hole, around the tree and back down the hole" is the most familiar method of tying a bowline. Most amateurs are knot-challenged so it's a great advantage to know at least this one.

INSTALLING TOWER SECTIONS

The easiest way to erect a tower is to use a crane. It's fast and safe but more expensive than doing it in sections by hand. To erect a tower by sections, a gin-pole is needed (see Fig 28.13). It consists of two pieces — a clamp or some device to attach it to the tower leg and a pole with a pulley at the top. The pole is typically longer than the work piece being hoisted, allowing it to be held above the tower top while being attached or manipulated.

With the gin-pole mounted on the tower, the haul rope runs up from the ground, through the gin-pole mast and pulley at the top of the gin-pole, and back down the tower. The haul rope has a knot (preferably a bowline) on the end for attaching things to be hauled up or down. A carabiner hooked into the bight of the knot can be attached to objects quickly so that you don't have to untie and re-tie the bowline with every use.

It's a good idea to pass the haul rope through a snatch-block at the bottom of the tower. This changes the direction of the rope from vertical to horizontal, allowing the ground crew to stand away from the tower (and the fall zone for things dropped off the tower) to manipulate the haul rope while also watching what's going on up on the tower.

GUYS

For guyed towers, an important construction parameter is guy wire material and *guy tension*. Do *not* use rope or any other material not rated for use as guy cable as permanent tower guys. Guyed towers for amateurs typically use either ³⁄₁₆-inch or ¼-inch *EHS* (extra high strength) steel guy cable. The only other acceptable guy material is Phillystran — a lightweight cable made of Kevlar fibers. Phillystran is available with he same breaking strength as EHS cable. The advantage of Phillystran is that it is non-conducting and does not create unwanted electrical in-

Table 28.2
Rope types and safe working loads in pounds.
Ropes are three-strand twisted unless otherwise noted.

	Diameter (inches)			
	1/4	3/8	1/2	5/8
Manila	120	215	425	700
Nylon	180	400	700	1150
Polypropylene	210	450	710	1055
Nylon braid-on-braid	420	960	1630	2800
Dacron braid-on-braid	350	750	1400	2400

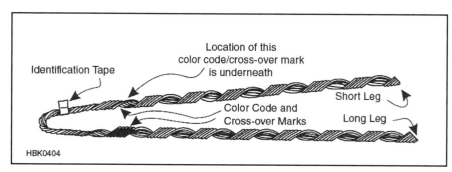

Fig 28.16 — A PreFormed Line Products Big Grip for guy wires.

teraction with antennas on the tower. It is an excellent choice for towers supporting multiple Yagi and wire antennas and does not have to be broken up into short lengths with insulators.

EHS wire is very stiff — to cut it, use a hand-grinder with thin cutting blades or a circular saw with a pipe-cutting aggregate blade. Be sure to wear safety glasses and gloves when cutting since there will be lots of sparks of burning steel being thrown off. Phillystran can be cut with a utility knife or a hot knife for cutting plastic.

If the guys are too loose, the result will be wind-induced shock loading. Guys that are too tight exert extra compressive load on the tower legs, reducing the overall capacity and reliability of the tower. The proper tension of EHS or Phillystran guys is 10% of the material's *ultimate breaking strength*. For 3/16-inch EHS the ultimate breaking strength is 4900 pounds and for 1/4-inch it's 6000 pounds so the respective guy tension should be 490 pounds and 600 pounds. The easiest to use, most accurate, and least expensive way to measure guy tension is by using a Loos tension gauge. It was developed for sailboat rigging so it's available at some marine supply stores or from Champion Radio Products.

Guy wires used to be terminated in a loop with cable clamps but those have been largely replaced by pre-formed Big Grips, shown in **Fig 28.16**. These simply twist onto the guy wire and are very secure. They grip the guy cable by squeezing the cable as tension is applied. Be sure to use the right type of Big Grips for the thickness and material of the guy cable.

28.2.4 Antenna Installation

Now that the tower is up, install the antennas. VHF/UHF whips and wire antennas are pretty straightforward, but installing an HF Yagi is a more challenging proposition. With a self-supporting tower, there are no guy wires to contend with — generally, the antenna can just be hauled up the tower face. Sometimes it is that easy!

In most cases, short of hiring a crane, the easiest way to get a Yagi up and down a tower is to use the *tram* method. A single tramline is suspended from the tower to the ground and the load is suspended under the tramline. Another technique is the *trolley* method in which two lines are suspended from the tower

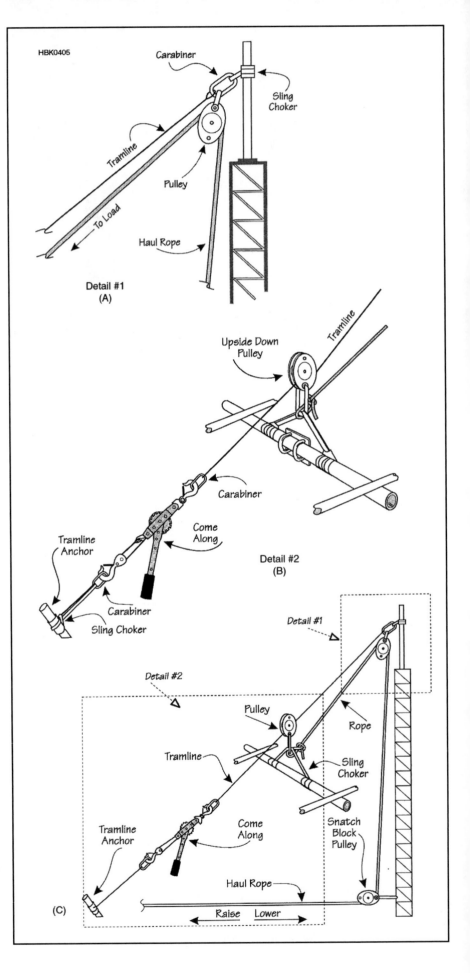

Fig 28.17 — At A, rigging the top of the tower for tramming antennas. Note the use of a sling and carabiner. (B) Rigging the anchor of the tramline. A come-along is used to tension the tramline. (C) The tram system for getting antennas up and down. Run the antenna part way up the tramline for testing before installation. It just takes a couple of minutes to run an antenna up or down once the tramline is rigged.

Table 28.3
Yield Strengths of Mast Materials

Material	Specification	Yield Strength (lb/in.^{2})
Drawn aluminum tube	6063-T5	15,000
	6063-T832	35,000
	6061-T6	35,000
	6063-T835	40,000
	2024-T3	42,000
Aluminum pipe	6063-T6	25,000
	6061-T6	35,000
Extruded alum. tube	7075-T6	70,000
Aluminum sheet and plate	3003-H14	17,000
	5052-H32	22,000
	6061-T6	35,000
Structural steel	A36	33,000
Carbon steel, cold drawn	1016	50,000
	1022	58,000
	1027	70,000
	1041	87,000
	1144	90,000
Alloy steel	2330 cold drawn	119,000
	4130 cold worked	75,000
	4340 1550 °F quench 1000 °F temper	162,000
Stainless steel	AISI 405 cold worked	70,000
	AISI 440C heat-treated	275,000

(From *Physical Design of Yagi Antennas* by David B. Leeson, W6NL)

to the ground and the antenna rides on top of the lines like a trolley car on tracks. Problems with the trolley technique include trying to get the lines to have the same tension, balancing the antenna so that it won't fall off of the lines, and the added friction of pulling the antenna up two lines. The tram method has none of these problems. **Fig 28.17** illustrates the tram method of raising antennas.

Tram and trolley lines are typically attached to the mast above the top of the tower. In the case of a big load, the lines may exert enough force to bend the mast. If in doubt, *back-guy* the mast with another line in the opposite direction for added support.

MASTS

A mast is a pipe that sticks out of the top of the tower and connects the rotator to the antenna. For small antenna loads and moderate wind speeds, any pipe will work. But as wind speed and wind load increase, more force will be exerted on the mast.

There are two materials used for masts — *pipe* and *tubing*. Pipe can be water pipe or conduit (EMT). Pipe is a heavy material with not much strength since its job is just to carry water or wires. Pipe is acceptable as mast material for small loads only. Another problem is that 1.5-in. pipe (pipe is measured by its inside diameter or ID) is only 1.9-in. OD. Since most antenna boom-to-mast hardware is designed for a 2-in. mast, the less-than-perfect fit may lead to slippage.

For any larger load use carbon-alloy steel tubing rated for high strength. A moderate antenna installation in an 80 MPH wind might exert 40,000 to 50,000 pounds per square inch (psi) on the mast. Pipe has a yield strength of about 35,000 psi, so you can see that pipe is not adequately rated for this type of use. Chromoly steel tubing is available with yield strengths from 40,000 psi up to 115,000 psi but it is expensive. **Table 28.3** shows the ratings of several materials used as masts for amateur radio antennas.

Calculating the required mast strength can be done by using a software program such as the *Mast, Antenna and Rotator Calculator (MARC)* software. (See the References.) The software requires as inputs the local wind speed, antenna wind load, and placement on the mast. The software then calculates the mast bending moment and will recommend a suitable mast material.

28.2.5 Weatherproofing Cable and Connectors

The biggest mistake amateurs make with coaxial cable is improper weatherproofing. (Coax selection is covered in the chapter on **Transmission Lines**.) **Fig 28.18** shows how to do it properly. First, use high-quality electrical tape, such as 3M Scotch 33+ or Scotch 88. Avoid inexpensive utility tape. After tightening the connector (use pliers carefully to seat threaded connectors — hand-tight isn't good enough), apply two wraps of tape around the joint.

When you're done, sever the tape with a knife or tear it very carefully — *do not* stretch the tape until it breaks. This invariably leads to "flagging" in which the end of the tape loosens and blows around in the wind. Then let the tape relax before finishing the wrap.

Next put a layer of butyl rubber *vapor wrap* over the joint. (This tape is also available in the electrical section of the hardware store.) Finally, add two more layers of tape over the vapor wrap, creating a professional-quality joint that will never leak. Finally, if the coax is vertical, be sure to wrap the final layer so that the tape is going *up* the cable as shown in Fig 28.18. In that way, the layers will act like roofing shingles, shedding water off the connection. Wrapping it top to bottom will guide water between the layers of tape.

28.2.6 Climbing Safety

Tower climbing is a potentially dangerous activity, so you'll need to use the proper safety equipment and techniques. OSHA, the Federal Occupational Safety and Health Administration, publishes rules for workplace safety. Although amateurs are not bound by those rules, you'll be much better off by following them.

First, if you are still climbing with a waist-only safety belt and leather positioning lanyard, throw them away! They are illegal and dangerous. The most important piece of climbing equipment is a *fall arrest harness* (FAH). Along with the waist safety belt and D-rings, it also has adjustable suspenders, leg loops and a D-ring between the shoulder blades for a *fall arrest (FA) lanyard*. **Fig 28.19** shows how to wear the harness for tower climbing.

A tower climber must be attached to the tower 100% of the time. One method of attachment is the *waist-positioning lanyard* shown in **Fig 28.20**. **Fig 28.21** shows a climber properly positioned in the harness on the tower with the positioning lanyard holding the climber in working position. The lanyard can be fixed-length or an adjustable rope or webbing type. Fixed-length varieties are usually too short or too long for optimum placement, but they are the least expensive. All fall-arrest equipment complies with OSHA rules, so what is actually used depends mostly on personal choice and budget.

There are two choices for FA lanyards. One is fixed-length and the other is *shock-absorbing*. A falling person generates a lot of force in a short time so keep the FA lanyard attached above you with enough slack to allow movement around the tower. That minimizes the distance of any fall. The stitched loops of a shock-absorbing FA lanyard yield gradually, decelerating to avoid a sudden stop at the end of a fall. The shock-absorbing lanyard costs more than a fixed-length lanyard.

Suitably equipped, start climbing the tow-

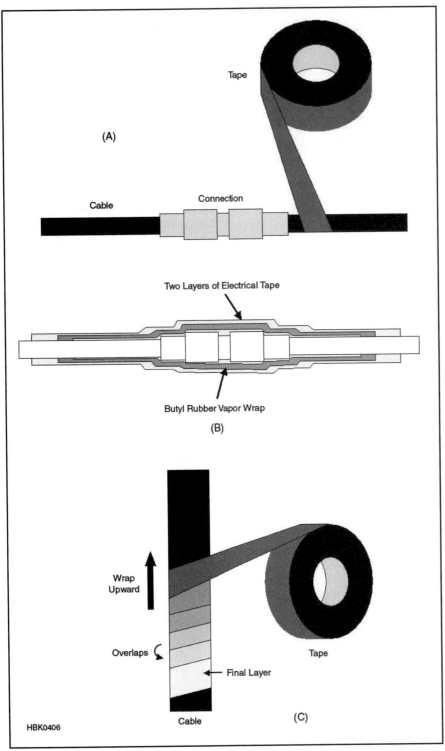

Fig 28.18 — Waterproofing a connector in three steps. At A, cover the connectors with a layer of good-quality electrical tape. B shows a layer of butyl rubber vapor wrap between the two layers of electrical tape. C shows how to wrap tape on a vertical cable so that the tape sheds water away from the connection. (Drawing (C) reprinted courtesy of *Circuitbuilding for Dummies*, Wiley Press)

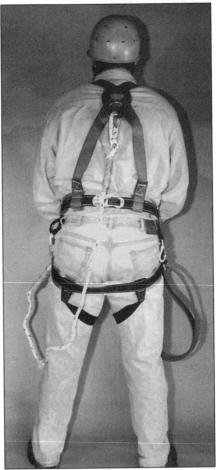

Fig 28.19 — (A) The well-dressed tower climber. Note the waist D-rings for positioning lanyard attachment as well as the suspenders and leg loops. At (B) is an adjustable positioning lanyard. The climber also has working boots, gloves, safety glasses and hardhat. (*K7LXC photos*)

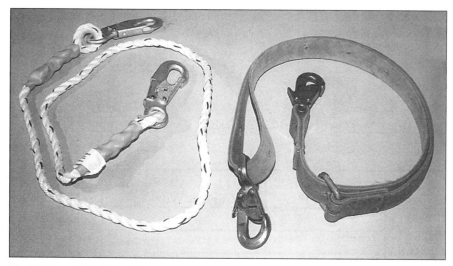

Fig 28.20 — A fixed-length rope positioning lanyard on the left and a versatile adjustable lanyard on the right. (*K7LXC photos*)

er. In order to be attached to the tower 100% of the time, attach the FA lanyard as high above you as you can and then climb up to it. Attach the positioning lanyard, detach the FA lanyard, and attach it above you again.

The positioning lanyard can also be moved up as you climb. As you reach the FA lanyard attachment point, unhook it, reposition, and repeat as many times as necessary. There are several varieties of FA lanyards including models with dual lanyards, thus allowing the lanyards to be alternated during climbing.

GROUND CREW SAFETY

The climber on the tower is the boss. Before tower work starts, have a safety meeting with the ground crew. Explain what is going to be done and how to do it as well as introducing them to any piece of hardware with which they may not be familiar (for example, carabiners, slings or come-along winches).

As part of the ground crew, there are a few rules to follow:

1) The climber on the tower is in charge.
2) Don't do anything unless directed by the climber in charge on the tower. This includes handling ropes, tidying up, moving hardware, and so on.
3) If not using radios to communicate, when talking to the climber on the tower, look up and talk directly to him or her in a loud voice. The ambient noise level is higher up on the tower because of traffic, wind and nearby equipment.
4) Communicate with the climber on the tower. Let him or her know when you're ready or if you're standing by or if there is a delay. Advise the climber when lunch is ready!

28.2.7 Antenna and Tower Safety References

AN Wireless — **www.anwireless.com**
ARRL Volunteer Counsel program — **www.arrl.org/FandES/field/regulations/local/vci.html**
ARRL Volunteer Consulting Engineer program — **www.arrl.org/FandES/field/regulations/local/vcei.html**
Brede, D., W3AS, "The Care and Feeding of an Amateur's Favorite Antenna Support — the Tree," *QST*, Sep 1989, pp 26-28, 40.
Champion Radio Products — **www.championradio.com**
Hopengarten, F., K1VR, *Antenna Zoning for the Radio Amateur* (ARRL, 2002)
Knot-tying Web site — **www.animatedknots.com**
Loos tension gauge — **www.championradio.com/rigging.html**

Fig 28.21 — The fall-arrest lanyard is above the climber so that the climber can climb up to it. The fall-arrest and positioning lanyards are then "leapfrogged" so that the climber remains attached to the tower 100 percent of the time. (*K7LXC photos*)

MARC software — **www.championradio.com/misc.html**
Morris, S., K7LXC, *Up The Tower: The Complete Guide To Tower Construction* (Champion Radio Products, 2009)
Rohn Tower — **www.rohnnet.com/rohn-catalog**
Straw, R.D., N6BV, ed., *The ARRL Antenna Book*, 21st ed. (ARRL, 2007)
Trylon — **www.trylon.com**
US Towers — **www.ustower.com**

Field Day Log

page _____ of _____

Call Used _____ Exchange Sent _____

Frequency	Mode	Date	Time (UTC)	Station Worked	Exchange Received

Field Day Log

page _____ of _____

Call Used _____   Exchange Sent _____

Frequency	Mode	Date	Time (UTC)	Station Worked	Exchange Received

Field Day Log

page _____ of _____

Call Used _____ Exchange Sent _____

Frequency	Mode	Date	Time (UTC)	Station Worked	Exchange Received

Field Day Log

page _____ of _____

Call Used _____   Exchange Sent _____

Frequency	Mode	Date	Time (UTC)	Station Worked	Exchange Received

Field Day Log

page _____ of _____

Call Used _____ Exchange Sent _____

Frequency	Mode	Date	Time (UTC)	Station Worked	Exchange Received

Field Day Log

page _____ of _____

Call Used _____ Exchange Sent _____

Frequency	Mode	Date	Time (UTC)	Station Worked	Exchange Received

Field Day Log

page _____ of _____

Call Used _____ Exchange Sent _____

Frequency	Mode	Date	Time (UTC)	Station Worked	Exchange Received

Field Day Log

page _____ of _____

Call Used _____   Exchange Sent _____

Frequency	Mode	Date	Time (UTC)	Station Worked	Exchange Received

Field Day Log

page _____ of _____

Call Used _____ Exchange Sent _____

Frequency	Mode	Date	Time (UTC)	Station Worked	Exchange Received

Field Day Log

page _____ of _____

Call Used _____   Exchange Sent _____

Frequency	Mode	Date	Time (UTC)	Station Worked	Exchange Received

Field Day Log

page _____ of _____

Call Used _____   Exchange Sent _____

Frequency	Mode	Date	Time (UTC)	Station Worked	Exchange Received

Field Day Log

page _____ of _____

Call Used _____ Exchange Sent _____

Frequency	Mode	Date	Time (UTC)	Station Worked	Exchange Received

Field Day Log

page _____ of _____

Call Used _____   Exchange Sent _____

Frequency	Mode	Date	Time (UTC)	Station Worked	Exchange Received

Field Day Log

page _____ of _____

Call Used _____ Exchange Sent _____

Frequency	Mode	Date	Time (UTC)	Station Worked	Exchange Received

Field Day Log

page _____ of _____

Call Used _____   Exchange Sent _____

Frequency	Mode	Date	Time (UTC)	Station Worked	Exchange Received

Field Day Log

page _____ of _____

Call Used _____   Exchange Sent _____

Frequency	Mode	Date	Time (UTC)	Station Worked	Exchange Received

Field Day Log

page _____ of _____

Call Used _____   Exchange Sent _____

Frequency	Mode	Date	Time (UTC)	Station Worked	Exchange Received

Field Day Log

page _____ of _____

Call Used _____ Exchange Sent _____

Frequency	Mode	Date	Time (UTC)	Station Worked	Exchange Received

Field Day Log

page _____ of _____

Call Used _____ Exchange Sent _____

Frequency	Mode	Date	Time (UTC)	Station Worked	Exchange Received

Field Day Log

page _____ of _____

Call Used _____ Exchange Sent _____

Frequency	Mode	Date	Time (UTC)	Station Worked	Exchange Received

Field Day Log

page _____ of _____

Call Used _____   Exchange Sent _____

Frequency	Mode	Date	Time (UTC)	Station Worked	Exchange Received

Field Day Log

page _____ of _____

Call Used _____ Exchange Sent _____

Frequency	Mode	Date	Time (UTC)	Station Worked	Exchange Received

Field Day Log

page _____ of _____

Call Used _____   Exchange Sent _____

Frequency	Mode	Date	Time (UTC)	Station Worked	Exchange Received

Field Day Log

page _____ of _____

Call Used _____   Exchange Sent _____

Frequency	Mode	Date	Time (UTC)	Station Worked	Exchange Received

Field Day Log

page _____ of _____

Call Used _____   Exchange Sent _____

Frequency	Mode	Date	Time (UTC)	Station Worked	Exchange Received

Field Day Log

page _____ of _____

Call Used _____ Exchange Sent _____

Frequency	Mode	Date	Time (UTC)	Station Worked	Exchange Received

Field Day Log

page _____ of _____

Call Used _____   Exchange Sent _____

Frequency	Mode	Date	Time (UTC)	Station Worked	Exchange Received

Field Day Log

page _____ of _____

Call Used _____   Exchange Sent _____

Frequency	Mode	Date	Time (UTC)	Station Worked	Exchange Received

Field Day Log

page _____ of _____

Call Used _____   Exchange Sent _____

Frequency	Mode	Date	Time (UTC)	Station Worked	Exchange Received

FEEDBACK

Please use this form to give us your comments on this book and what you'd like to see in future editions, or e-mail us at **pubsfdbk@arrl.org** (publications feedback). If you use e-mail, please include your name, call sign, e-mail address and the book title, edition and printing in the body of your message. Also indicate whether or not you are an ARRL member.

Where did you purchase this book? ☐ From ARRL directly ☐ From an ARRL dealer
Is there a dealer who carries ARRL publications within:
 ☐ 5 miles ☐ 15 miles ☐ 30 miles of your location? ☐ Not sure.

 License class:
 ☐ Novice ☐ Technician ☐ Technician with code ☐ General ☐ Advanced ☐ Amateur Extra

Name _____ ARRL member? ☐ Yes ☐ No
_____ Call Sign _____
Address _____
City, State/Province, ZIP/Postal Code _____
Daytime Phone ( ) _____ Age _____
If licensed, how long? _____
Other hobbies _____ E-mail _____

Occupation _____

For ARRL use only	FD HB
Edition	2 3 4 5 6 7 8 9 10 11 12
Printing	1 2 3 4 5 6 7 8 9 10 11 12

From _____
_____
_____

Please affix postage. Post Office will not deliver without postage.

EDITOR, FIELD DAY HANDBOOK
ARRL—THE NATIONAL ASSOCIATION FOR AMATEUR RADIO
225 MAIN STREET
NEWINGTON CT 06111-1494

please fold and tape